GROW
YOUR OWN
MEDICINE

Edible Healing Plants In your Garden

GROW
YOUR OWN
MEDICINE

Edible Healing Plants In your Garden

Koh Hwee Ling

Glenis Lim Yi Jing

Marcus Wong Xiu Ren

Tan Chay Hoon

National University of Singapore

World Scientific

NEW JERSEY • LONDON • SINGAPORE • BEIJING • SHANGHAI • HONG KONG • TAIPEI • CHENNAI • TOKYO

Published by

World Scientific Publishing Co. Pte. Ltd.

5 Toh Tuck Link, Singapore 596224

USA office: 27 Warren Street, Suite 401-402, Hackensack, NJ 07601

UK office: 57 Shelton Street, Covent Garden, London WC2H 9HE

British Library Cataloguing-in-Publication Data
A catalogue record for this book is available from the British Library.

GROW YOUR OWN MEDICINE
Edible Healing Plants in your Garden

ISBN 978-981-12-8786-2 (hardcover)
ISBN 978-981-12-8803-6 (paperback)
ISBN 978-981-12-8787-9 (ebook for institutions)
ISBN 978-981-12-8788-6 (ebook for individuals)

For any available supplementary material, please visit
https://www.worldscientific.com/worldscibooks/10.1142/13720#t=suppl

Authors and Contributors

Authors

Koh Hwee Ling

Glenis Lim Yi Jing

Marcus Wong Xiu Ren

Tan Chay Hoon

Other Contributors

Poh Keng Ling

Tay Zhi Wen

He Ronghui

Kang Li Jia

Kang Li Hui

Neo Soek Ying

Disclaimer

The contents of the book serve to provide both general and scientific information about medicinal plants, their uses and planting them in your home garden. The information is not intended as a guide to self-medication by consumers or to treatment by healthcare professionals. The general public is advised to discuss the information contained herein with a physician, pharmacist, nurse or other authorised healthcare professionals. Neither the authors nor the publisher can be held responsible for the accuracy of the information itself or the consequences from the use or misuse of the information in this book.

The resources are not vetted, and it is the reader's responsibility to ensure the accuracy of the information cited. Readers are reminded that the information presented is subject to change as research is ongoing, and there may be inter-individual variations. While every effort is made to minimise errors, there may be inadvertent omissions or human errors in compiling these monographs.

Guide to Using This Book

If you are curious about the useful medicinal plants from nature, wondering what food may be useful to health, and which edible medicinal plant you can grow in your garden, this book is for you! The book is intended for both the general public as well as healthcare professionals. Hence, as much as possible, the original terminologies from the information source is used to avoid any miscommunication.

This book is a compilation of a total of 30 medicinal plants. Most of these plants are also eaten as food.

For each plant, information on scientific name, common name, origin, phyto-constituents, medicinal uses, pharmacological activities, clinical trials (if any), dosage (if any), adverse reactions, toxicity, contraindications (who should not use), precautions, drug-herb interactions (if any), uses as food, growth conditions and cultivation (e.g. light, water and soil preferences) and harvesting is collated. Authors' notes are added where appropriate.

Names: The plants are listed in alphabetical order according to their scientific names. The family of the plant is shown in brackets. Some common names including Chinese names with Chinese phonetics (where appropriate) are shown below the scientific names.

Phytoconstituents: These are the reported components present in the plant. Some of them may be the active components responsible for the activities.

Medicinal uses: These refer to the traditional uses. Wherever available, information from official sources (e.g. The Chinese Pharmacopoeia, German Commission E) will be shown.

Pharmacological activities: These refer to the scientifically studied and reported biological effects of extracts of the plants or of phytoconstituents associated with the plant. These may be from *in vitro* studies or *in vivo* studies in animals.

Clinical trials: Clinical trials are scientific studies to test the effects of the medicinal plants using human volunteers. Unlike pharmaceuticals (drugs), where clinical trials are mandatory, clinical trials on medicinal plants are not common or are not available at all. Selected information on clinical trials of the relevant plants, if any, is shown.

Dosage refers to the amount used. This is often available only when an official monograph (e.g. Chinese Pharmacopoeia, German Commission E) is available and accessible.

Safety: Although medicinal plants are naturally occurring, it does not mean that they are totally safe. Hence, safety related information, wherever available, will be presented. This includes *adverse reactions, toxicity, contraindications* (who should not use), *precautions, drug-herb interactions* (if any). Adverse reactions are undesirable effects associated with the plant or with its components that have been reported. Reports of toxic effects in animals or humans are presented under "Toxicity". They may be scientific studies typically in animals or other models to study whether the plant or its phytoconstituents have any toxic effects. Contraindications and precautions are presented together as they are closely related. Drug-herb interactions refer to effects of using both drug and plant together. The plant or plant phytoconstituents may increase or decrease the activity of the drug, or may affect the absorption, distribution, metabolism or elimination of the drug, or may add on to the drug effects or have an opposite effect to that of the drug. The human CYP450 enzymes in the liver (e.g. CYP1A2, CYP3A4) help to metabolise drugs. Selected enzymes that are known to be affected by the plant or phytoconstituents may be shown in this section.

Uses as food: Most of the medicinal plants are also consumed as food and related information is presented. Some are eaten raw, cooked and taken as beverages, etc.

Growth conditions: For readers who are interested to "grow their own medicine", information on the sunlight, water and soil required for planting are collated.

Harvesting: The plant parts to be harvested and the estimated time/period for harvesting are mentioned.

Authors' notes: This will be included when there is interesting information that may not fall nicely in the previous categories.

References: The number of references for each plant is intentionally kept small, up to a maximum of 6 for each plant. They are selected based on reliability, relevance and year of publication.

Acknowledgement

This book is possible with the support of the following:

Lee Foundation [A-000-284-500-00] for financial support

Madam Lee Ying for her passion and support of medicinal plant research

Dr Wilson Wong, NParks, for valuable and stimulating discussions

Uncle Tan Thean Teng for his generous and unfailing sharing of medicinal plant knowledge and plant samples

Mr Ng Kim Chuan for his generous sharing of medicinal plant samples

Our families for their unwavering support

Contents

Plant Monographs

Abelmoschus esculentus **(L.) Moench** (Malvaceae)

Lady's Finger, Okra
羊角豆 (Yang Jiao Dou)

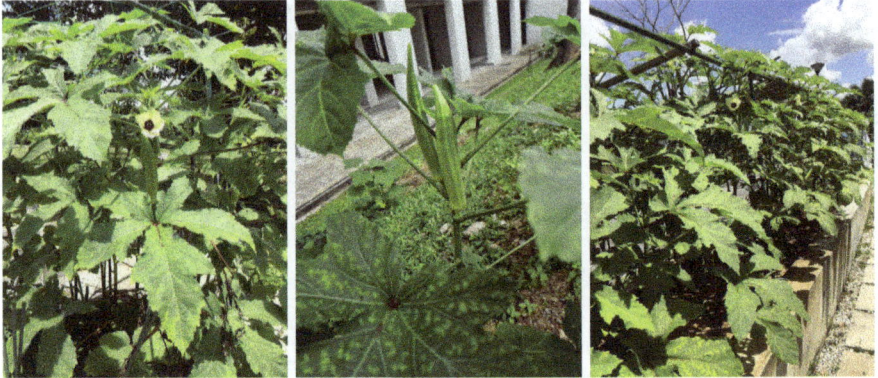

Origin

Native to tropical Africa and cultivated in India, Egypt, Turkey, Brazil and Greece

Phytoconstituents

Polyphenols, catechins, polysaccharides, flavonoids, quercetin, isoquercitrin, quercetin 3-O-glucoside, quercetin-3-O-gentiobiose, galacturonic acid and others

Medicinal Uses

<Fruits>: Relieve inflammation, diuretic and emollient. Used for anaemia, cough, dysentery, dysuria, gonorrhoea and sore throat

Pharmacological Activities

Anti-Alzheimer's, anti-cancer, anti-diabetic, anti-hyperlipidaemic, anti-inflammatory, anti-obesity, anti-oxidant, cognitive enhancement, hepatoprotective, immunomodulatory, neuroprotective and wound healing

Clinical Trials

<Okra powder>:
- Significantly improved fasting blood sugar levels, triglyceride, total cholesterol and LDL cholesterol in type 2 diabetes.

Dosage

No information as yet.

Adverse Reactions

No information as yet.

Toxicity

No information as yet.

Contraindications/Precautions

No information as yet.

Drug-Herb Interactions

<*A. esculentus* with metformin>: Vasoprotective effect from attenuation of high fat induced endothelial inflammation.

<Sulfated *A. esculentus* extract>: Inhibit CYP1A and glutathione-S-transferase *in vivo*, suggesting its chemopreventive potential.

Uses as Food

The immature fresh or dried fruit, and sometimes the leaves and seeds, are consumed as food. The tender green fruits can be eaten raw or used as a vegetable in cooking.

Growth Conditions

Direct sunlight for 6–8 hours per day. Moderate water. Well-draining soil, fertile loamy soil.

Harvesting

The tender green pods can be harvested in 47–97 days from first sowing. The fruit may take between 55 and 65 days to grow.

Aloe vera **(L.) Burm. f.** (Liliaceae)

Aloe, Lidah Buaya
芦荟 (Lu Hui)

Origin

Native to northern, southern and eastern Africa

Phytoconstituents

Aloin, aloe-emodin, aloenin, aloesin, emodin and others

Medicinal Uses

<Fresh gel>: Emollient and topical remedy for minor skin conditions such as sunburn, rashes, burns, wounds, abrasions, bruises and swelling. Applied on the forehead as headache remedy and on the hair to promote healthy hair growth

<Dried leaf juice>: Used for constipation. According to the Chinese Pharmacopoeia, it is applied externally for sore and fungal infections. According to the Ayurvedic Pharmacopoeia of India, it is used for ascites, dysmenorrhea, fever and liver disorders

4

Pharmacological Activities

Anti-bacterial, anti-cancer, anti-diabetic, anti-inflammatory, anti-oxidant, anti-periodontitis, anti-ulcer, improve skin elasticity, improve ulcerative colitis and wound healing

Clinical Trials

<Topical gel>:
- Significant improvements in bone defect depth, defect depth reduction, probing depth, clinical attachment levels for patients with chronic periodontitis.
- Significantly improved wound healing for mothers who underwent a caesarean section
- Significant decrease in ulcer size, pain, erythema and exudation for patients with recurrent aphthous stomatitis.
- Prevention of pressure ulcers for patients.
- Prevention of traumatic oral ulceration for patients with fixed orthodontic appliances.

<Dried *A. vera* mesophyll powder tablets>:
- Significantly improved skin elasticity for healthy individuals.

Dosage

<125–500 mg powdered dried leaf juice>: Refer to medicinal uses in Chinese Pharmacopoeia.

<0.04–0.11 g dried leaf juice>: Laxative.

Adverse Reactions

Abdominal cramps and pain are the most common adverse events associated with oral use of the latex.

Toxicity

<Overdose or long-term use of *A. vera* latex>: Hepatitis, hypokalaemia, albuminuria, haematuria and severe diarrhoea with consequent fluid and electrolyte imbalance.

<Aloe-emodin>: Induce liver cell injury and hepatotoxicity.

<8 weeks of oral administration at 150–300 mg/kg of ethanol extract of *A. vera* gel>: Anti-fertility effects in rats.

<*A. vera* green rind extract if consumed for long periods at high doses>: Significant increase in creatinine levels, resulting in possible kidney toxicity in rats.

Contraindications/Precautions

<Dried leaf latex or juice>: Contraindicated in intestinal stenosis, ileus, atony, severe dehydration with electrolyte loss, chronic constipation, nephritis, haemorrhoids, appendicitis, colic, Crohn's disease, cramps, ulcerative colitis, irritable bowel syndrome, diverticulitis and idiopathic abdominal pain.

<Gel>: Contraindicated in people with known allergies to plants in the Liliaceae family.

Drug-Herb Interactions

<Dried leaf latex or juice with loop or thiazide diuretics, corticosteroids or liquorice root>: May potentiate risk for hypokalaemia.

<Leaf juice>: Shorten intestinal transit time, decreasing absorption of orally administered drugs.

<Emodin>: Inhibition of CYP1A2, CYP2C9, CYP2C19, CYP2D6 and CYP3A4 enzymes.

<Aloe-emodin>: Strongly inhibited CYP1A2 and CYP2C19 enzymes.

Uses as Food

A. vera gel harvested from leaves is edible and can be incorporated into food such as beverages, confectionery and dairy products.

Growth Conditions

Full sun, direct sunlight for 6–8 hours per day. Little water. Well-draining, porous, dry, pH balanced soil and in cactus and succulent potting soil mix.

Harvesting

Harvest matured leaves typically after 3 to 4 years. After cutting the leaf, allow the yellow latex to drain out completely. To get the aloe vera gel (colourless and gelatinous), obtain a fresh, washed leaf and slice away the rind. Wash under water to remove any yellow latex.

Alternanthera sessilis (L.) R.Br. ex DC.
(Amaranthaceae)

Sessile Joyweed
红田乌草 (Hong Tian Wu Cao)

Origin

Originated in Asia and is native to Singapore

Phytoconstituents

Phenolic acids, flavonoids, fatty acids, diterpenes, triterpenes, carotenoids and others

Medicinal Uses

Fever, diarrhoea, leprosy, skin disorders, wounds, agalactia, dyspepsia, burning sensations, haemorrhoids and splenomegaly

Pharmacological Activities

Analgesic, anti-allergic, anti-bacterial, anti-cataract, anti-diabetic, anti-diarrhoeal, anti-fungal, anti-hypertensive, anti-inflammatory, anti-oxidant, hypolipidaemic and wound healing

Clinical Trials

No information as yet.

Dosage

<30 g decoction of whole plant>: Fever, dysentery, constipation, jaundice, peptic ulcer, sore throat, urinary tract infection, toothache, tuberculosis and intestinal inflammation.

Adverse Reactions

No information as yet.

Toxicity

No information as yet.

Contraindications/Precautions

No information as yet.

Drug-Herb Interactions

No information as yet.

Uses as Food

In Asia, *A. sessilis* is commonly used as a vegetable. It can be eaten raw as a salad or cooked as herbal vegetable either stir-fried or in soup. The leaves can also be boiled in water to prepare herbal tea.

Growth Conditions

Full sun, direct sunlight for 6–8 hours per day. Moderate water. Fertile loamy, well-draining soil.

Harvesting

The shoots, leaves and flowers can be harvested when required.

Authors' Notes

A. sessilis is available as the green and red varieties.

Andrographis paniculata (Burm. f.) Nees
(Acanthaceae)

Bitterweed, King of Bitters, Common Andrographis
穿心莲 (Chuan Xin Lian), 苦草 (Ku Cao)

Origin

Originates from Southern to Southeast Asia, including China and India

Phytoconstituents

Andrographolides, andrographidines, andrograpanin, andrographon and others

Medicinal Uses

<Plant>: Liver and blood disorders, diabetes, worm infestation, skin diseases, fever in malaria and general physical weakness

<Dried aerial parts>: According to the Chinese Pharmacopoeia, it is indicated for sore throat, sores on the tongue and mouth, whooping cough, dysentery, diarrhoea, common cold associated with fever, painful heat strangury, swollen abscesses and snake, worm or insect wounds

<Leaves>: Prophylactic and symptomatic treatment of uncomplicated respiratory tract infections such as common cold, sore throat, cough, sinusitis, bronchitis and to relieve fever. Topically, a leaf poultice is applied to treat skin diseases, boils, snake bites, leprosy, ulcers, swelling and pruritus

Pharmacological Activities

Analgesic, anti-arthritic, anti-bacterial, anti-cancer, anti-diabetic, anti-fibrotic, anti-inflammatory, anti-oxidant, anti-viral, anti-SARS-CoV-2, cardioprotective, hepatoprotective, immunomodulatory, nephroprotective, neuroprotective, ulcerative colitis treatment and upper respiratory tract infection prevention/treatment

Clinical Trials

<*A. paniculata* extracts containing andrographolide>:
- Significantly reduce cold symptoms; pain and joint swelling in rheumatoid arthritis; fatigue in multiple sclerosis and triglyceride levels in hypertriglyceridaemia.
- Significantly increase clinical remissions and responses in ulcerative colitis.

Dosage

<6–9g of dried aerial parts>: Refer to medicinal uses in Chinese Pharmacopoeia.

Adverse Reactions

<Various *A. paniculata* preparations>: Headache, nausea, mild skin rash, urticaria, abdominal pain, diarrhoea, fatigue, thirst and vomiting.

<Andrographolide>: Fatigue, allergic reaction, headache, rash, diarrhoea, nausea, metallic or bitter taste, tastelessness, tongue dryness, reduced sex drive, photosensitivity of eyes, reduced short term memory, lymphadenopathy, tender lymph nodes, heartburn and dizziness.

<Andrographolide derivative injections>: Gastrointestinal disorders, skin and subcutaneous tissue disorders, and anaphylaxis.

Toxicity

In various studies regarding *A. paniculata* extracts and andrographolide, the majority indicated that they were safe and nontoxic, except for the following:

- <Oral administration of 50 mg/kg/day andrographolide for 48 days>: Adversely affected spermatogenesis and testicular structure in male rats.
- <Oral administration of 1000 mg/kg *A. paniculata* leaf powder in water for 8 weeks>: Reduced sperm count, motility and viability, impaired spermatogenesis and significantly reduced circulatory testosterone in male rats, inducing reproductive toxicity.

Contraindications/Precautions

Use of the dried aerial parts is contraindicated in people with known allergies to plants of the Acanthaceae family. Not recommended for use during pregnancy or lactation due to abortifacient effect.

Drug-Herb Interactions

<Aqueous and methanol extracts of *A. paniculata*>: Inhibition of CYP3A4 enzyme.

<Andrographolide>: Inhibition of UGT2B7 enzyme.

<*A. paniculata* extract and andrographolide with naproxen>: Reduced exposure of naproxen in rats.

<Andrographolide with naproxen>: Synergistic anti-arthritic activity in rats.

<*A. paniculata* extract with cisplatin and 5-fluorouracil>: Enhanced anti-tumour and anti-metastatic effects in mice.

<Andrographolide with anti-PD-1 antibody>: Improved function of CD4+ and CD8+ T cells and enhanced anti-tumour effect of anti-PD-1 in a mouse model of colorectal cancer.

<Andrographolide sulfonate with imipenem>: Significantly increased survival rate of mice with acute pneumonia.

<*A. paniculata* with dihydroartemisinin-piperaquine (DHP)>: Lower foetal morphologic abnormalities, reduced DHP toxicity in pregnant mice infected with malaria.

<*A. paniculata* extract with ribavirin>: Severity of adverse drug reactions increased.

<*A. paniculata* methanolic extract with polyvalent anti-snake venom for Naja naja venom>: More effective in neutralizing the toxicity of Naja naja venom.

Uses as Food

A. paniculata is not normally consumed as food, but is consumed as tea.

Growth Conditions

Full sun with direct sunlight for 6–8 hours per day. Moderate water. Well-draining soil, fertile loamy soil.

Harvesting

The aerial parts of the plant are ready for harvest when flowers are in bloom. Flowers may start appearing 3–5 months from sowing.

Azadirachta indica A. Juss. (Meliaceae)

Neem Tree, Indian Lilac, Margosa Tree
印度苦楝 (Yin Du Ku Lian)

Origin

Originates from India and Java, commonly found in South and Southeast Asia

Phytoconstituents

Azadirachtin, azadirone, azadiradione, nimbolide, nimbidin, nimbin, nimbinin, nimbocinol, gedunin and others

Medicinal Uses

<Stem bark>: Anti-pyretic and anti-malarial

<Leaves>: Anti-pyretic, anti-septic and diuretic. Used for malarial fever, skin diseases, dental and gastrointestinal disorders, headache, heartburn, diabetes, ulcers and eczema

<Berries>: Laxative and emollient

<Seed oil>: Insecticide and skin disorders

Pharmacological Activities

Anti-apoptotic, anti-bacterial, anti-cancer, anti-diabetic, anti-hyperlipidaemic, anti-inflammatory, anti-malarial, anti-oxidant, anti-pyretic, anti-viral, chemo-preventive, immunomodulatory, mosquito repellent and neuroprotective

Clinical Trials

<Bark extract powder capsules>:
- Significantly decreased gastric acid secretion and healed ulcers.

<Gel containing *A. indica* leaf extract>:
- Effective in inhibiting plaque growth.

<Aqueous extract of *A. indica* leaves and twigs (PhytoBGS®) capsules>:
- Significantly improved blood sugar parameters, endothelial function and reduced oxidative stress, systemic inflammation in type 2 diabetes mellitus.
- Significantly decreased oxidative stress, inflammation and insulin resistance, improved glucose tolerance and vascular function in endothelial dysfunction.

Dosage

<70% of ethanol extract of dried leaves diluted to 40%, applied externally twice a day>: Treatment of ringworms.

Adverse Reactions

The plant is deemed generally safe without adverse effects except for the following:
- Extremely rare contact allergy reported in a few cases.
- One report of ventricular fibrillation and cardiac arrest due to neem leaf poisoning.
- Anti-fertility effects of *A. indica* methanolic leaf and fruit extracts in rats.

Toxicity

<Oral administration of 50 mg/kg *A. indica* stem bark ethanolic extract for 21 days>: Significantly increased the weights of liver, kidney, lungs and heart in rats.

Contraindications/Precautions

Not recommended during pregnancy, nursing or children under the age of 12 due to potential genotoxic effects.

Drug-Herb Interactions

<*A. indica* with chloroquine sulphate>: Significantly decreased serum concentration, slowed absorption and elimination, and increased half-life.

<*A. indica* aqueous leaf extract and glipizide>: Lower hypoglycaemic effect in diabetic rats due to decreased bioavailability of glipizide.

<*A. indica* crude extracts>: Inhibition of CYP3A4, CYP1A2, CYP2C8, CYP2C9 and CYP2D6 enzymes *in vitro*.

Uses as Food

The leaves can be eaten raw or cooked. Neem leaf tea bags are also available commercially.

Growth Conditions

Full sun with direct sunlight for 6–8 hours per day. Moderate water. Dry, well-draining soil, fertile loamy soil.

Harvesting

The tree grows slowly at its initial stage but should start growing rapidly once it is one year old. Harvest should begin after its initial growth stages. Fruits are oblong and are initially green, turning yellow or purple when ripe.

Carica papaya L. (Caricaceae)

Papaya, Papaw
木瓜 (Mu Gua)

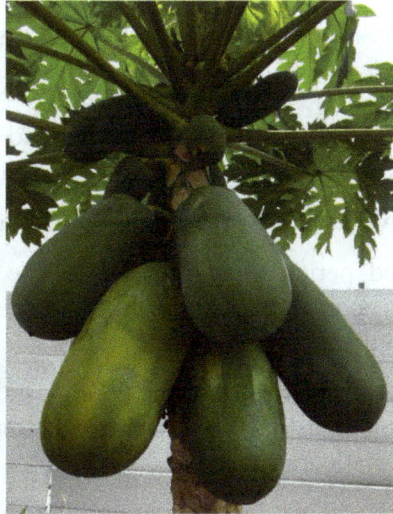

Origin

Native to Central America and is cultivated in tropical and sub-tropical regions such as China, India and Malaysia

Phytoconstituents

<Leaves>: Carpaine, methyl gallate, alkaloids, flavonoids and others

<Latex>: Papain, chymopapains and others

<Fruits>: Carpaine, carotenoids, chymopapains, lycopene, papain, quercetin, kaempferol, rutin and others

<Seeds>: Carpaine, benzyl isothiocyanate and others

Medicinal Uses

<Dried roots>: According to the Ayurvedic Pharmacopoeia of India, they are used for haemorrhoids, stones, ageusia, worm infestation, abdominal pain, gout, ulcers, menorrhagia, metrorrhagia, as well as skin, bleeding and urinary disorders

<Bark>: Toothache and abdominal disorders

<Leaves>: Malaria and related symptoms

<Latex>: Used topically for wound-healing, wound inflammation, corns, warts, eczema, psoriasis, freckles, snake bites, leprosy and haemorrhoids

<Flowers>: Anti-pyretic and emmenagogue

<Dried pericarp of ripe and unripe fruits>: According to the Ayurvedic Pharmacopoeia of India, they are used for cough, elimination of worms, blood disorders, gout and asthma

<Ripe fruit>: In Traditional Chinese Medicine, it is used to treat stomach-ache, indigestion, insufficient lactation, limb numbness, ulcers, eczema and intestinal parasites

<Seeds>: Treat insect bites and diarrhoea

Pharmacological Activities

Anthelmintic, anti-bacterial, anti-cancer, anti-diabetic, anti-inflammatory, anti-malarial, anti-oxidant, anti-thrombocytopenic, immunomodulatory, nephroprotective and wound healing

Clinical Trials

<*C. papaya* leaf extract tablets, leaf juice extract>:
- Significantly increased mean platelet count in dengue fever-associated thrombocytopenia.

<*C. papaya* leaf extract mouthwash, fermented papaya gel, dentifrice>:
- Effective in reducing interdental gingival bleeding and inflammation for healthy individuals.

<Dressing containing matured *C. papaya* fruit>:
- Significantly faster healing time and shorter hospitalisation length for post-caesarean section wound dehiscence patients.

Dosage

<10–20 g powdered dried pericarp of ripe and unripe fruits>: Refer to medicinal uses in Ayurvedic Pharmacopoeia of India.

<2–6 g powdered dried roots>: Refer to medicinal uses in Ayurvedic Pharmacopoeia of India.

Adverse Reactions

<Oral consumption of *C. papaya* leaf extract>: Rash, nausea and vomiting.

<Ingestion of papain>: Oesophageal perforation, oesophagitis and aspiration pneumonitis, in some cases leading to death.

<*C. papaya* seed extract>: Anti-fertility effects in female rats; anti-fertility activity due to negative effects on sperm motility parameters in men.

Toxicity

<Oral administration of 800 mg/kg of aqueous *C. papaya* seed extract for 10 days post-coitum>: Abortifacient and teratogenic effects in rats.

Contraindications/Precautions

Use of unripe fruits and papaya seeds during pregnancy should be avoided as *C. papaya* has been suggested to be abortifacient and teratogenic. Use during lactation should also be avoided due to insufficient information available. Normal consumption of ripe fruit as food should be fine.

Drug-Herb Interactions

<*C. papaya* dried leaf decoction with artesunate>: Significant anti-malarial effect in mice.

<*C. papaya* leaf ethanolic extract with metformin or glimepiride>: Reduced hypoglycaemic effects at the initial 2 hours and enhanced hypoglycaemic effect at 24 hours.

<Dried *C. papaya* leaf decoction of aqueous extract with digoxin>: Inhibition of p-glycoprotein transport of digoxin *in vitro in* Caco-2 cell monolayers, impeding the absorption and bioavailability of digoxin.

<Dried *C. papaya* leaf decoction with artemisinin>: Subsynergistic anti-malarial effect in inhibiting growth of *Plasmodium falciparum*.

<Fresh *C. papaya* leaf crude aqueous extract with artesunic acid>: Antagonistic anti-malarial effect in *Plasmodium berghei* infected mice.

<Freeze dried crude *C. papaya* leaf extract with ciprofloxacin>: Reduced absorption and serum half-life of ciprofloxacin in rabbits.

Uses as Food

The unripe fruit is consumed as vegetable while the ripe fruit is consumed as a common fruit or processed into jams, juice and preserved papaya. Seeds are not eaten as food.

Growth Conditions

Full sun with direct sunlight for 6–8 hours per day. Moderate water. Well-draining soil, fertile loamy soil, moist soil.

Harvesting

Papaya tree should start flowering 5–6 months after the seedlings have sprouted and fruits should appear shortly after, bearing fruit throughout the year. Young fruits are green in colour while mature fruits are orange and have a smooth texture.

Centella asiatica (**L.**) **Urban** (Apiaceae)

Asiatic Pennywort, Indian Pennywort, Gotu Kola
积雪草 (Ji Xue Cao)

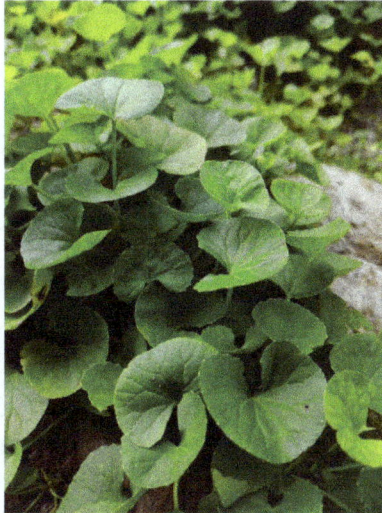

Origin

Native to warm regions including South and Central America, Africa, Australia, China, India, Singapore, Malaysia, Cambodia, Laos, Thailand, Vietnam, Indonesia, Madagascar and the Pacific Islands

Phytoconstituents

Asiaticoside, asiatic acid, madecassoside, madecassic acid and others

Medicinal Uses

<Dried herb>: According to the Chinese Pharmacopoeia, it is indicated for blood and stone strangury, diarrhoea due to summer heat, fall injuries, abscesses and sores

<Dried whole plant>: According to the Ayurvedic Pharmacopoeia of India, it is used for bleeding disorders, skin diseases, polyuria, fever, asthma, cough, tastelessness, anaemia, inflammation and itching

<Folk uses>: Syphilis, eye diseases, hypertension, wound-cleaning, haemorrhoids, dysentery, bronchitis, oedema, venous disorders, anaemia, cellulite, measles, cholera, constipation, dysmenorrhoea, smallpox and toothache

Pharmacological Activities

Anti-Alzheimer's, anti-arthritic, anti-bacterial, anti-cancer, anti-inflammatory, anti-oxidant, anxiolytic, cardioprotective, cognitive enhancement, hepatoprotective, neuroprotective, neurotrophic, treatment of venous disorders and wound healing

Clinical Trials

<Total triterpenic fraction of *C. asiatica*, a standardized extract in tablet>:
- Significantly reduced oedema, resting flux (blood flow), venoarteriolar response and rate of ankle swelling for patients in diabetic microangiopathy; reduced flux at rest and rate of ankle swelling for patients with venous hypertension.

<*C. asiatica* extract capsule>:
- Significantly improved calmness, alertness, as well as accuracy and speed of working memory.

<*C. asiatica* extract gel>:
- Improved skin erythema and wound appearance for acne scars.

<*C. asiatica* ceramide cream>:
- Improved skin barrier hydration in skin dryness.

Dosage

<15–30 g of dried herb>: Refer to medicinal uses in Chinese Pharmacopoeia

<3–6 g of dried whole plant>: Refer to medicinal uses in Ayurvedic Pharmacopoeia of India.

Adverse Reactions

<Oral consumption of *C. asiatica*>: Usual doses may result in headache, nausea, dyspepsia and stomach discomfort, while an overdose may result in drowsiness, dizziness and transient unconsciousness; increase blood glucose and lipid levels.

Toxicity

Acute toxicity studies involving oral administration of a standardised extract of *C. asiatica* to rats have shown that the lethal dose is greater than 2000 mg/kg, and sub-chronic toxicity studies also showed that no significant toxicity was observed at doses of up to 1000 mg/kg.

<Consumption of *C. asiatica* for 20 to 60 days>: Case reports of hepatotoxicity in humans.

<Oral administration of aqueous *C. asiatica* leaf extract for 60 days>: Reduced sperm count and testicular damage, suggesting anti-fertility effects in male rats.

Contraindications/Precautions

Contraindicated in people with hypersensitivity to plants from the Apiaceae family. Due to insufficient information available, usage of the herb during pregnancy, lactation and in paediatrics is not recommended. Caution should be taken in long-term topical application of the herb, as it could be linked to increased risk of neoplasm.

Drug-Herb Interactions

As high oral doses may cause drowsiness, sedative medications should not be taken concurrently. May potentially interfere with diabetes and hyperlipidaemia treatment, as its ability to alter blood glucose and lipid levels has been postulated.

<*C. asiatica* hydroalcoholic extract with valproate or phenytoin>: Improved anti-epileptic efficacy. Significant increase in oral bioavailability and reduction

in clearance, as well as an increase in rate of absorption and decrease in volume of distribution in rats.

Uses as Food

Commonly consumed as vegetable or as beverage in many countries.

Growth Conditions

Semi-shade, direct sunlight for 4–6 hours. Lots of water. Moist soil, waterlogged soil. Sandy soil, clay soil.

Harvesting

C. asiatica matures in 90 days. The leaves and stems are edible and can be harvested for use as food. It is advisable to wear gloves when harvesting as touching the leaves could cause skin irritation.

Authors' Notes

A component of *C. asiatica,* asiaticoside, is available commercially as a cream or ointment for wound healing and scars.

Clitoria ternatea **L.** (Leguminosae)

Butterfly Pea, Blue Pea
蝴蝶花豆 (Hu Die Hua Dou)

Origin

Originates in South America or the Malay Archipelago and can be found throughout the tropics

Phytoconstituents

Taraxerol, clitorienolactones, ternatins, kaempferol, quercetin, anthocyanins, flavonoids and others

Medicinal Uses

<Roots>: Laxative, diuretic and cathartic effects. Used to treat ascites and fever

<Leaves>: Treat acne, pustules and swollen joints

<Flower extract>: Eye inflammation

Pharmacological Activities

Anti-Alzheimer's, anti-bacterial, anti-depressant, anti-diabetic, anti-fungal, anti-hypertensive, anti-inflammatory, anti-oxidant, anti-pyretic, enhance memory, neuroprotective and nootropic

Clinical Trials

<*C. ternatea* flower extract in water>:
- Significantly improved postprandial glucose, insulin and anti-oxidant status in the presence of sucrose. In a fasting state, significantly increases plasma anti-oxidants without causing hypoglycaemia.

Dosage

No information as yet.

Adverse Reactions

No information as yet.

Toxicity

<Oral administration of *C. ternatea* aerial and roots ethanolic extracts above 2000 mg/kg>: Lethargy and droopy eyelid in mice.

<Intraperitoneal administration of *C. ternatea* aerial and roots ethanolic extracts at 2900 mg/kg and above>: Severe Central Nervous System depression and death in rats.

Contraindications/Precautions

No information as yet.

Drug-Herb Interactions

No information as yet.

Uses as Food

The flowers, tender fruits, young shoots and leaves are eaten as vegetable. Dried butterfly pea flowers are commonly used to brew tea, found in tea blends and are commercially available. The blue flowers are also used as a natural food colouring for rice, traditional kueh and drinks.

Growth Conditions

Full sun with direct sunlight for 6–8 hours per day. Moderate water. Well-draining soil, fertile loamy soil.

Harvesting

The plant takes around 12 weeks to become established, and flowering should begin 6–8 weeks after planting. Harvest the flowers when in bloom.

Cymbopogon citratus **(DC.) Stapf** (Poaceae)

Lemongrass, Citronella Grass
香茅 (Xiang Mao)

Origin

Originates in Southeast Asia

Phytoconstituents

Citral, citronellal, limonene, geraniol, neral, myrcene, β-pinene, linalool, cis-verbenol, nerol and others

Medicinal Uses

<*C. citratus* and its oil>: Mild astringent, gastrointestinal discomfort and disorders, muscle pain and neuralgia, colds, nervous disturbances, exhaustion, malaria treatment and as a calmative

Pharmacological Activities

Anaesthetic, anthelmintic, anti-bacterial, anti-cancer, anti-carcinogenic, anti-diabetic, anti-fungal, anti-inflammatory, anti-oxidant, anti-tumour, gastroprotective, insecticidal, mosquito repellent and neuroprotective

Clinical Trials

<*C. citratus* essential oil in olive oil>:
- 98.8% bite protection from mosquitoes, making it a suitable repellent.

<*C. citratus* essential oil by inhalation>:
- Anxiolytic effect by improving psychological and physical parameters.

Dosage

<40–60 g, decoction or infusion drunk 3 times daily>: Dysuria without swelling problems.

Adverse Reactions

Toxic alveolitis due to inhalation.

Toxicity

Various toxicity studies involving *C. citratus* and its extracts indicate that they were nontoxic, except for the following:

<*C. citratus* essential oil up to a concentration of 1000 µg/mL>: Highly toxic to brine shrimp larvae.

<*C. citratus* essential oil above 0.02 mg/mL after 48 hours treatment>: Cardiotoxicity and shortened tail in zebrafish.

Contraindications/Precautions

No information as yet.

Drug-Herb Interactions

<*C. citratus* ethanolic extract with taxol or mitoxantrone>: Significantly increased induction of apoptosis in DU-145 prostate cancer cells.

<*C. citratus* ethanolic extract with taxol or combination treatment of 5-fluorouracil, folinic acid and oxaliplatin>: Did not inhibit apoptosis induction by chemotherapeutics in colorectal cancer cells.

<*C. citratus* ethanolic extract with combination treatment of folinic acid, 5-fluorouracil and oxaliplatin>: Mitigation of adverse effects from the combination treatment and improve chemotherapeutic efficacy in colorectal cancer in mice.

Uses as Food

Lemongrass is used in many Southeast Asian dishes and is an essential ingredient for a variety of food in Thailand, Vietnam, and Cambodia.

Growth Conditions

Light: Full sun with direct sunlight for 6–8 hours per day. Moderate water. Well-draining soil, moist soil, loamy soil, able to tolerate most soil types.

Harvesting

Lemongrass takes 4–6 months to grow and can be harvested when the stalk has grown to about 30 cm in height.

Gynura procumbens **(Lour.) Merr.** (Asteraceae)

Longevity Spinach, Sambung nyawa
尖尾凤 (Jian Wei Feng)
蛇接骨 (She Jie Gu)

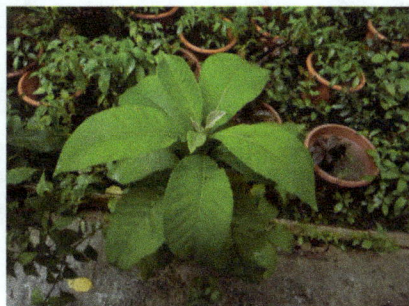

Origin

Grows in Borneo, Java, Philippines and Malaysia, native to Singapore

Phytoconstituents

Chlorogenic acid, polyphenols, flavonoids, terpenoids and others

Medicinal Uses

<Roots>: Renal disorders and diarrhoea

<Leaves>: Tumour, inflammation, high cholesterol, diabetes, high blood pressure and cancer

Pharmacological Activities

Anti-bacterial, anti-cancer, anti-diabetic, anti-hyperlipidaemic, anti-hypertensive, anti-inflammatory, anti-oxidant, anti-proliferative, anti-pyretic, cardioprotective and vasorelaxant

Clinical Trials

<*G. procumbens* extract herbal gel>:

- Significantly reduces the infection incidence and has an anti-inflammatory effect in recurrent cold sores.

Dosage

No information as yet.

Adverse Reactions

Allergic contact dermatitis.

Toxicity

Various toxicity studies have indicated that *G. procumbens* extracts do not result in toxicity or death.

Contraindications/Precautions

No information as yet.

Drug-Herb Interactions

<Aqueous, methanolic and ethanolic *G. procumbens* leaf extracts>: Inhibit human liver glucuronidation activity *in vitro*.

Uses as Food

Young leaves can be eaten raw or cooked.

Growth Conditions

Semi-shade with direct sunlight for 4–6 hours per day. Moderate water. Well-draining soil and moist soil.

Harvesting

Leaves and stems may be harvested within 1 month from planting.

Hibiscus sabdariffa **L.** (Malvaceae)

Hibiscus, Sour Tea, Jamaica Sorrel
洛神葵 (Luo Shen Kui)

Origin

Native to tropical Africa and Asia, including India and Malaysia

Phytoconstituents

Anthocyanins, quercetin, kaempferol, myricetin, catechin, hibiscus acid, chlorogenic acid, organic acids and others

Medicinal Uses

<Dried roots>: According to the Ayurvedic Pharmacopoeia of India, they are used in chronic diarrhoea, throat diseases, bone fracture and ulcers

<Calyces>: Hypertension, anaemia, fever, liver and bilious disorders, urinary calculi, cancer, heart diseases and dysuria

<Seeds>: Demulcent, tonic and diuretic, used for dyspepsia, dysuria, strangury and physical weakness

Pharmacological Activities

Anti-apoptotic, anti-atherosclerotic, anti-bacterial, anti-cancer, anti-hyperlipidaemic, anti-hypertensive, anti-inflammatory, anti-oxidant, cardio-protective, cognitive enhancement, diuretic, hepatoprotective, nephroprotective and neuroprotective

Clinical Trials

<*H. sabdarrifa* tea>:
- Significantly reduced mean arterial pressure, systolic blood pressure and diastolic blood pressure but increased urine volume and creatinine clearance in hypertension.

<Capsules containing aqueous extract of *H. sabdarrifa*>:
- Significantly decreased serum triglycerides, body weight, body mass index and low density lipoprotein in hyperlipidaemia.

Dosage

<5–10 g of dried roots>: Refer to medicinal uses in Ayurvedic Pharmacopoeia of India.

Adverse Reactions

<Oral use of *H. sabdariffa*>: Abdominal distention, epigastric pain, flatulence, headache and blurred vision.

Toxicity

<Oral consumption of 20 g/L dried *H. sabdariffa* petals infusion containing 0.5 mg/L of aluminium for 16 days>: Increased urinary aluminium in humans.

<Sub-chronic oral administration of up to 500 mg/kg *H. sabdariffa* calyces aqueous extract for 90 days in rats>: Decrease in direct bilirubin, which suggest compromised liver excretory function, alteration of serum electrolytes, which suggest compromised kidney function and significant decrease in platelet concentration in rats.

Contraindications/Precautions

Should not be used in pregnancy as it is claimed to be an emmenagogue and may be an abortifacient; should not be used during lactation due to insufficient evidence available.

Drug-Herb Interactions

<Aqueous *H. sabdariffa* calyces extract with simvastatin>: Significantly decreased total cholesterol and triglyceride levels in rats.

<Aqueous *H. sabdariffa* calyces extract with captopril>: Reduced efficacy of captopril in rats.

<Aqueous *H. sabdariffa* calyces extract with glibenclamide>: More favourable treatment outcomes for diabetes mellitus and associated oxidative stress in rats.

<Aqueous *H. sabdariffa* calyces extract>: Weakly inhibit CYP1A2, CYP2C8, CYP2D6, CYP2B6, CYP2C19, CYP3A4, CYP2C9 and CYP2A6 enzymes.

<*H. sabdariffa* calyces methanol extract with hydrochlorothiazide>: Significantly increased urine volume in rats and plasma concentration of the drug in rabbits.

<*H. sabdariffa* aqueous decoction with amlodipine>: Better anti-hypertensive effect in rats.

<Aqueous beverage containing *H. sabdariffa* calyces with simvastatin>: Significantly reduced plasma concentration and peak concentration of simvastatin in healthy human volunteers.

Uses as Food

Calyces are used to prepare drinks, jams, jellies, wines and sauces, and is also a natural food colourant due to its anthocyanin content.

Growth Conditions

Direct sunlight for 6–8 hours per day. Moderate water. Well-draining soil, moist soil.

Harvesting

Roselle plant takes 4–6 months to mature. The calyces are harvested between 7 and 10 days after the flowers have bloomed.

Impatiens balsamina **L.** (Balsaminaceae)

Balsam
凤仙花 (Feng Xian Hua)

Origin

Native to India and is commonly found in southern Asia as a garden plant

Phytoconstituents

2-methoxy-1,4-naphthoquinone, balsaminones, balsamina sterol, anthocyanins, α-amyrin, β-sitosterol, apigenin derivatives and others

Medicinal Uses

<Folk medicine>: Rheumatism, isthmus, generalised pain, fractures, fingernail inflammation, scurvy, carbuncle, dysentery, bruises and beriberi

<Flowers>: Lumbago, neuralgia, burns and scalds

<Dried ripe seeds>: According to the Chinese Pharmacopoeia, in Traditional Chinese Medicine, they are used for amenorrhoea, abnormal growth or swelling in abdomen and dysphagia occlusion

Pharmacological Activities

Anti-bacterial, anti-cancer, anti-diabetic, anti-inflammatory, anti-microbial, anti-neurodegenerative, anti-nociceptive, anti-oxidant and anti-pruritic

Clinical Trials

No information as yet.

Dosage

<3–5 g of seeds decocted with water taken orally>: Refer to medicinal uses in Chinese Pharmacopoeia.

Adverse Reactions

No information as yet.

Toxicity

<5 or 10 mg/mL hydroalcoholic extracts of *I. balsamina* stems>: Significantly reduced roundworm survival, resulted in neuronal toxicity and had negative impacts on reproduction.

Contraindications/Precautions

Use with caution during pregnancy.

Drug-Herb Interactions

No information as yet.

Uses as Food

The leaves, young shoots, petals and seeds can be eaten cooked.

Growth Conditions

Full sun with direct sunlight for 6–8 hours per day. Moderate water. Fertile loamy soil, moist soil, well-draining soil.

Harvesting

Balsam takes at least 60 days to mature and flower.

Ipomoea batatas (**L.**) **Lam.** (Convolvulaceae)

Sweet Potato, Keledek
甘薯 (Gan Shu)

Origin

Originates from tropical America

Phytoconstituents

Anthocyanins, polyphenols, quercetin, kaempferol, β-carotene, β-sitosterol, folic acid, caffeic acid and others

Medicinal Uses

<Roots>: Tonic for the stomach, spleen and kidneys in Chinese medicine

<Leaves>: Anti-diabetic, anti-inflammatory, anti-bacterial and anti-anaemic

Pharmacological Activities

Anti-ageing, anti-atherosclerotic, anti-cancer, anti-diabetic, anti-hyperlipidaemic, anti-inflammatory, anti-obesity, anti-oxidant, anti-tumour, gut microbiota modulation, prebiotic, hepatoprotective and immunomodulatory

Clinical Trials

<Purple fleshed sweet potato beverage>:
- Significantly lower hepatic biomarkers GGT, AST and ALT, indicating hepatoprotective effects in borderline hepatitis.

<Purple sweet potato leaves>:
- Significant increase in proliferation responsiveness of the immune system.

Dosage

No information as yet.

Adverse Reactions

Sporamin A protein has been found to be an allergen.

Toxicity

Heavy metal levels detected were below toxicological reference values and therefore present no toxicological risk to consumers. Toxicological tests on mice resulted in no obvious toxicity.

Contraindications/Precautions

No information as yet.

Drug-Herb Interactions

No information as yet.

Uses as Food

The tubers, leaves and young shoots are cooked and eaten.

Growth Conditions

Full sun with direct sunlight for 6–8 hours per day. Moderate water. Fertile loamy soil, well-draining soil, dry soil.

Harvesting

Depending on its propagation method and part of plant to be harvested, harvest time will vary. Harvest of the leaves and stems may begin 1 week after planting pieces of sweet potato in soil.

Mentha spicata **L.** (Lamiaceae)

Spearmint, Common Mint, Bo He
留兰香 (Liu Lan Xiang)

Origin

Native to Europe and Asia

Phytoconstituents

Menthol, menthone, carvone, limonene, pinene, β-caryophyllene, 1,8-cineole, rosmarinic acid, polyphenols, flavonoids and others

Medicinal Uses

Common cold, cough, sinusitis, fever, bronchitis, nausea, vomiting, indigestion, intestinal colic, loss of appetite, gastrointestinal and respiratory disorders, bad breath, dandruff, and as a diuretic and sedative agent

Pharmacological Activities

Anti-bacterial, anti-cancer, anti-cholinesterase, anti-diabetic, anti-fungal, anti-inflammatory, anti-obesity, anti-oxidant, anti-proliferative, hepatoprotective, insect repellent, insecticidal and neuroprotective

Clinical Trials

<Spearmint extract capsules>:
- Significantly improved working memory in age-associated memory impairment.

<Spearmint tea>:
- Significant decrease in androgenic hormones in hirsutism.

<High rosmarinic acid spearmint tea>:
- Significantly improved stiffness, physical disability and pain scores in knee osteoarthritis.

<Proprietary spearmint extract>:
- Significantly improved reactive agility, complex attention and sustained attention in young, active individuals.

Dosage

<A tablespoon of fresh spearmint leaves added to hot water>: Stomachache.

Adverse Reactions

Contact allergy to the leaves, contact cheilitis from its essential oil and anaemia have been reported. Dose-dependent hepato- and nephrotoxicity have also been reported in murine models. Heartburn may be worsened by spearmint tea.

Toxicity

<Oral doses of 20 g/L spearmint tea as drinking water over 30 days>: Significant increases in malondialdehyde levels, indicative of lipid peroxidation damage, as well as uterine damage and apoptosis in rats.

<Leaves and aerial parts of *M. spicata*>: Toxic digestive effects of intestinal gas and colon bloating.

Contraindications/Precautions

Avoid use in people with known allergies to spearmint or other members of the Lamiaceae family.

Drug-Herb Interactions

No information as yet.

Uses as Food

Spearmint is used as a spice in food, as well as flavouring in gums and confectionary.

Growth Conditions

Full sun with direct sunlight for 6–8 hours per day. Moderate water. Well-draining soil, fertile loamy soil, poor infertile soil, waterlogged soil.

Harvesting

Spearmint leaves are regularly harvested to promote growth of the plant. Leaves should be harvested before or after flowering for the best flavour.

Momordica charantia **L.** (Cucurbitaceae)

Bitter Gourd, Bitter Melon
苦瓜 (Ku Gua)

Origin

The origin of the plant is uncertain, but it is cultivated in the tropics and sub-tropical regions

Phytoconstituents

Momordicosides, momorcharins, momordicines, charantin, caffeic acid, cucurbitacins, karavilagenins, karavilosides, kuguacins, kuguaglycosides, polypeptide-p, saponins, triterpenoids, triterpene glycosides and others

Medicinal Uses

<Crushed leaves>: Used topically in skin disorders, scalds, burns, stomach ache, and as a poultice to alleviate headache

<Leaf extract>: Celiac disease, cough, fever, nausea and vomiting, and as an anthelmintic

<Unripe fruits>: Common cold with fever, heat stroke, body ache, dysentery, rheumatism, gout and liver and spleen disorders

<Fresh fruits>: Listed in the Ayurvedic Pharmacopoeia of India for use in skin diseases, urinary and blood disorders, jaundice, anaemia, fever, asthma, cough and loss of sense of taste

Pharmacological Activities

Anti-cancer, anti-diabetic, anti-hyperlipidaemic, anti-inflammatory, anti-obesity, anti-oxidant, cardioprotective, gastroprotective, hepatoprotective, immunomodulatory, increase insulin secretion, nephroprotective, neuroprotective and wound healing

Clinical Trials

<Capsules containing dried powdered *M. charantia* fruit pulp>:
* Significant reduction in HbA1c, 2-hour plasma glucose after oral glucose tolerance test, mean fructosamine levels, fasting plasma glucose, plasma sialic acid and triglyceride in type 2 diabetes.

* Significant increase in insulin secretion in type 2 diabetes.

<*M. charantia* fruit capsules>:
* Significant improvement in analgesic scores and reducing pain in knee osteoarthritis.

<*M. charantia* hot water extract capsules>:
* Significantly lowered LDL cholesterol levels.

Dosage

<10–15 mL of fresh fruit juice>: Skin diseases, urinary and blood disorders, jaundice, anaemia, fever, asthma, cough and loss of sense of taste.

Adverse Reactions

More commonly reported adverse events in clinical trials include headache, diarrhoea, dizziness and nausea.

Toxicity

<Petroleum ether, chloroform and ethanol *M. charantia* seed extracts intraperitoneally administered at 250 mg/kg for 48 days>: Antispermatogenic and androgenic activity in male mice.

<Oral administration of aqueous *M. charantia* leaf extract for 30 days>: Antifertility effects in terms of reductions of oestrogen and progesterone levels in female rats.

<Intraperitoneal administration of α- and β-momorcharin>: Terminating pregnancy in mice.

<Momorcharins from *M. charantia* seeds>: Teratogenic to mouse embryos *in vitro*, where head, limbs and trunk malformations were observed.

<Aqueous extracts of *M. charantia* of up to 1000 μg/mL>: Decreased survival and hatching rate of zebrafish embryos. Scoliosis and different heartbeats were observed in zebrafish larvae.

<*M. charantia* seed extract of 5–30 μg/mL>: Teratological effects in zebrafish embryos.

<*M. charantia* fruit extract of 30 μg/mL and above>: Cardiac hypertrophy in zebrafish embryos.

Contraindications/Precautions

Use of *M. charantia* is not recommended during pregnancy due to abortifacient and teratogenic effects. *M. charantia* should also not be used during lactation or by children due to case reports of hypoglycaemia and seizures in children.

Drug-Herb Interactions

<Various preparations of *M. charantia*>: Inhibit P-glycoprotein and CYP2C9, and modulate resistance to aminoglycosides and chemotherapeutic agents.

<*M. charantia* fruit extract with glibenclamide>: Greater improvement in glycaemic control than glibenclamide alone in streptozotocin-diabetic rats.

<*M. charantia* juice with gemcitabine>: Enhanced efficacy specifically against *KRAS*-mutant tumours.

Uses as Food

The unripe fruit is consumed as a common vegetable around the world and has considerable nutritional value.

Growth Conditions

Full sun with direct sunlight for 6–8 hours per day. Moderate water. Well-draining soil, fertile loamy soil.

Harvesting

Bitter gourd should begin to flower in about 2–3 months after sowing seeds in ground. Fruits should develop and be ready for harvest 2–3 weeks after flowers have bloomed. Unripe green fruits can be harvested in 5–6 months after sowing seeds in pots or containers, before the fruits begin to turn yellow.

Morinda citrifolia **L.** (Rubiaceae)

Noni, Cheese Fruit
海巴戟 (Hai Ba Ji)

Origin

Originates from Southeast Asia, India, Indochina and Australasia, and is widely distributed in Polynesia. It is a native plant in Singapore

Phytoconstituents

Iridoids, morindicone, morindone, morinthone, anthraquinones, ursolic acid, quercetin, rutin, β-sitosterol and others

Medicinal Uses

<Leaves>: Wounds, fever, gingivitis, pharyngitis, boils and breast inflammation

<Fruits>: Diarrhoea, dysentery, intestinal worms, cough, tuberculosis, eye complaints, fever with vomiting, gingivitis, sore throat, thrush, blood poisoning, leucorrhoea and skin abscesses

Pharmacological Activities

Anti-arthritic, anti-bacterial, anti-cancer, anti-diabetic, anti-fatigue, anti-fungal, anti-hyperlipidaemic, anti-inflammatory, ant-ioxidant, anxiolytic, immunomodulatory and neuroprotective

Clinical Trials

<Tahitian noni juice>:
- Significantly decreased plasma superoxide anion radicals and lipid hydroperoxide levels; lipid hydroperoxide- and malondialdehyde-DNA adduct levels; total cholesterol, triglycerides and homocysteine levels, and LDL levels in heavy smokers.

- Significantly increased HDL levels in heavy smokers; increased mean and maximal oxygen uptake in athletes.

Dosage

No official information as yet.

Adverse Reactions

<Tahitian noni juice>: Allergic reactions, diarrhoea, nausea and skin rash.

<Capsules containing dried *M. citrifolia* fruit>: Nausea and abdominal discomfort.

Toxicity

<Consumption of aqueous *M. citrifolia* fruit extract of up to 300 mg/kg and Tahitian noni juice of up to 20 mL/kg daily for 9 days>: Induced foetal bone defects without affecting maternal health in pregnant rats.

<*M. citrifolia* fruit aqueous extract of 400 mg/kg via intra-gastric gavage for 21 days>: Adverse effects to both pregnant rat and fetus, visceral and skeletal anomalies of fetuses, maternal hepatotoxicity and preimplantation embryonic losses in pregnant rats.

Contraindications/Precautions

Noni contains a high potassium content. Caution should be taken in patients who are hyperkaelimic or are required to restrict potassium intake, such as in kidney disease. Use in pregnancy, lactation and patients with liver disease is not recommended due to insufficient evidence available.

Drug-Herb Interactions

Concomitant use with angiotensin-converting enzyme inhibitors, angiotensin receptor blockers or potassium-sparing diuretics may potentiate hyper-kalaemia risk, due to the high potassium content of *M. citrifolia*.

<Aqueous *M. citrifolia* fruit extract with ranitidine>: Increase rate and extent of ranitidine absorption in healthy volunteers

Uses as Food

Noni leaves, fruits and seeds are used as food.

Growth Conditions

Full sun with direct sunlight for 6–8 hours per day. Moderate water. Well-draining soil, dry soil, waterlogged soil, poor infertile soil.

Harvesting

Noni may begin to fruit 9–12 months after planting. Unripe green fruits turn white as they grow and ripen. The fruits are harvested when they are ripening (beginning to turn white), or when they are fully ripe.

Moringa oleifera **Lam.** (Moringaceae)

Horseradish Tree, Drumstick Tree
辣木 (La Mu)

Origin

Originates from India and can be found throughout the tropics

Phytoconstituents

Moringine, moringinine, β-carotene, thiamine, riboflavin, niacin, quercetin, kaempferol, β-sitosterol, chlorogenic acid, amino acids and others

Medicinal Uses

<Roots>: Fever, asthma, dropsy, dyspepsia, otalgia, epilepsy, hysteria, gout, syphilis, rheumatism, spasms, hepatosplenomegaly, bladder and renal calculi

<Bark>: Abortifacient, anti-rheumatic, anti-scorbutic, diuretic, rubefacient and stimulant effects

<Leaves>: Gonorrhoea, syphilis, abdominal colic, dropsy, heal snake bites, dog bites, wounds and ulcers

Pharmacological Activities

Anti-bacterial, anti-cancer, anti-diabetic, anti-hyperlipidaemic, anti-inflammatory, anti-obesity, anti-oxidant, anti-tumour, anti-ulcer, chemopreventive, hepato-protective, nephroprotective, neuroprotective and wound healing

Clinical Trials

<Capsules of *M. oleifera* leaf ethanolic extract>:
- Significantly lowered body mass index, total cholesterol and LDL cholesterol levels in obesity.

<Capsules of dry *M. oleifera* leaf powder>:
- Decreased fasting blood glucose and HbA1c in prediabetes.

<*M. oleifera* leaves fried quickly in a minimal amount of vegetable oil for 10 minutes at 80°C>:
- Significantly decreased systolic and diastolic blood pressure.

<*M. oleifera* toothpaste>:
- Significant reduction in mean gingival index and plaque index scores.

Dosage

No information as yet.

Adverse Reactions

<*M. oleifera* leaves>: One report of Stevens–Johnson syndrome and one report of cutaneous toxicity.

<*M. oleifera* leaf extract supplement>: One report of sub-massive pulmonary embolism due to induced clot formation.

<*M. oleifera* methanol leaf extract>: Induced male reproductive toxicity shown by the significant decrease in serum testosterone, luteinising hormone and reduced sperm motility and count in rats.

<*M. oleifera* aqueous lipid-rich seed extract>: Unsafe during pregnancy, with observed renal and hepato toxicities in female rats.

Toxicity

<Oral doses of 2571 mg/kg *M. oleifera* seeds extract>: Gastrointestinal distention, stomach discolouration, increased liver mass, irregular respiratory patterns, piloerection and death in rats.

<Oral repeated doses of 150, 300 and 600 mg/kg *M. oleiferea* methanol leaf extract for 90 days>: Reduced body weight, cholesterol, and LDL but increased platelet count in rats.

Contraindications/Precautions

Caution is advised regarding consumption of *M. oleifera* seed extract by patients who are undergoing chemotherapy for breast cancer.

Drug-Herb Interactions

<*M. oleifera* leaf powder with amodiaquine>: Significantly decreased the Cmax of amodiaquine.

<Active fractions isolated from pods of *M. oleifera* with rifampicin>: Significantly increased rifampicin plasma concentration, Cmax, K, half-life, AUC and inhibited CYP450 enzyme activity.

<*M. oleifera* seed extract with doxorubicin and cyclophosphamide>: Upregulated several genes and pathways promoting cancer cell proliferation that are downregulated by chemotherapy in mice.

<*M. oleifera* methanol extract>: Inhibition of CYP1A2, CYP2C8, CYP2C9, CYP2C19, CYP2D6 and CYP3A4 enzyme activity.

Uses as Food

The flowers, leaves and seeds are all edible. The immature seed pods, called "Drumsticks", are used in cooking curry.

Growth Conditions

Full sun with direct sunlight for 6–8 hours per day. Moderate water. Fertile loamy soil, well-draining soil.

Harvesting

Horseradish Tree may mature in 8 months from sowing. The leaves, flowers, both immature and mature fruits can be harvested for use.

Authors' Notes

The leaves of the plant are nutritious with vitamins and minerals and the powder can be used for food fortification. Seed oil can be used in cosmetics and personal healthcare products.

Morus alba L. (Moraceae)

White Mulberry, Tut
桑 (Sang)

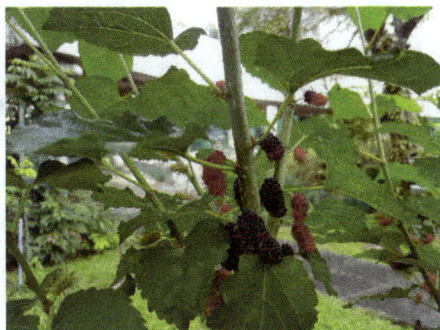

Origin

Originates from China, found from the North Western Himalayas to China

Phytoconstituents

Morusin, morusine, morusalnol, kuwanons, mulberranol, mulberroside A, quercetin, chlorogenic acid, anthocyanins and others

Medicinal Uses

<Leaves>: According to the Chinese Pharmacopoeia, they are used for common cold, dizziness and headache, red eyes and blurry vision. They are also used to purify the blood, treat gonorrhoea, influenza, headaches, epistaxis, sore throat, eye infection, hypertension, fever and lethargy.

<Fruits>: According to the Chinese Pharmacopoeia, they are used for dizziness and tinnitus, palpitation and insomnia and premature greying. They are also used for dizziness, lethargy, premature greying, anaemia, cough associated with tuberculosis, hypertension, urinary incontinence and tinnitus.

Pharmacological Activities

Anti-Alzheimer's, anti-bacterial, anti-cancer, anti-depressant, anti-diabetic, anti-hyperlipidaemic, anti-inflammatory, anti-obesity, anti-oxidant, anti-viral, cardioprotective and hepatoprotective

Clinical Trials

<Mulberry leaf extract containing 1-deoxynojirimycin (DNJ)>:
- Significantly reduces starch digestion and absorption when taken with meal, useful for postprandial glycaemic control.

<DNJ enriched powder mixed with water>:
- Significantly suppressed postprandial increase in glucose and insulin secretion.
- Significantly decreased fasting plasma glucose and HbA1c levels, with improvements in insulin resistance in borderline diabetes.

Dosage

<5–10 g dried leaves decocted with water, taken orally>: Refer to medicinal uses in Chinese Pharmacopoeia.

<9–15 g dried twigs decocted with water, taken orally>: Numbness and sore pain in the joint, shoulder and arm.

<6–12 g dried root bark decocted with water or made into powder, taken orally>: Cough and panting, oedema and low urine output.

<9–15 g dried fruit cluster decocted with water, taken orally>: Dizziness and tinnitus, palpitation and insomnia, and premature graying.

Adverse Reactions

No information as yet.

Toxicity

Majority of toxicity studies indicated that *M. alba* is safe and not genotoxic. However, a study on intraperitoneal doses of up to 2000 mg/kg *M. alba* ethanol leaf extract in mice resulted in biological, haematological and histological damage, particularly in the kidney, liver and spleen.

Contraindications/Precautions

<Dried root bark>: Contraindicated in lung deficiency, profuse urine and cough.

<Dried leaves>: Precaution with common cold, bland taste in mouth and cough with thin, white phlegm.

<Dried fruit cluster>: Contraindicated with thin, uniformed stool.

Drug-Herb Interactions

<Mulberroside A>: Significantly reduced P-glycoprotein expression and inhibited P-gylcoprotein, activated protein kinase C, nuclear factor kappa B and nuclear translocation.

<Mulberry leaves extract with metformin>: Enhanced anti-hyperglycaemic effect and increased systemic exposure of metformin.

Uses as Food

The fruit is eaten fresh, dried, or made into juice, jam and wine. The leaves are also edible and are used to brew Mulberry Leaf Tea.

Growth Conditions

Full sun with direct sunlight for 6–8 hours per day. Moderate water. Well-draining soil, clayey, loamy or sandy soil.

Harvesting

The plant bears green flowers and develops oblong fruits that are black, pink or white. When immature, the fruits change from white to pink, to red, and finally mature to black or dark purple in colour. Fully ripe fruits are harvested for use and consumption.

Murraya koenigii (L.) Spreng. (Rutaceae)

Curry Leaf Tree, Indian Curry Tree
麻绞叶 (Ma Jiao Ye)
咖喱叶 (Ga Li Ye)

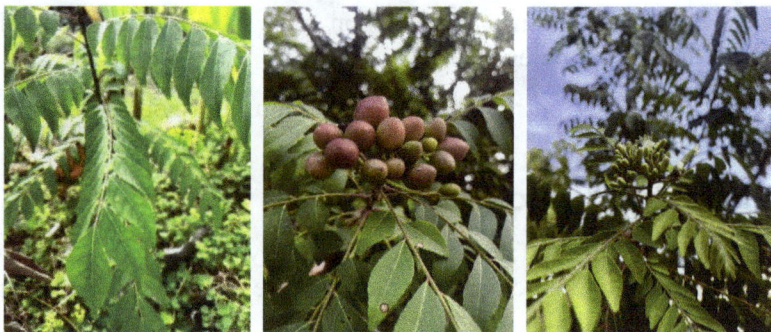

Origin

Native to India, Sri Lanka, Pakistan and other South Asian countries

Phytoconstituents

Mahanimbine, murrayakonines, girinimbine, koenimbine, koenine, kurrayam, mukonicine, isomahanimbine, bismahanine and others

Medicinal Uses

<Bark and roots>: Carminative and stimulant, and are used topically for bites and skin eruptions

<Dried leaves>: According to the Ayurvedic Pharmacopoeia of India, they are used for diarrhoea, vomiting, haemorrhoids, burning sensation, ulcer, fever, itching, helminthiasis, leprosy, skin diseases, urinary disorders, dysentery, oedema, colicky pain, cachexia and leukoderma.

Pharmacological Activities

Anti-cancer, anti-diabetic, anti-inflammatory, anti-oxidant, cardioprotective, hepatoprotective, nephroprotective, neuroprotective and wound healing

Clinical Trials

M. koenigii mouthwash was found to be equally as effective as commercially available chlorhexidine mouthwash.

Dosage

<3–6 g of powdered dried leaves, or 10–20 mL of juice from dried leaves>: Refer to medicinal uses in Ayurvedic Pharmacopoeia of India.

Adverse Reactions

No information as yet.

Toxicity

Majority of toxicity studies conducted indicate that chronic administration of *M. koenigii* leaf extracts are safe and non-toxic. However, one study indicated that a single dose oral administration of up to 2000 mg/kg *M. koenigii* methanolic leaf extract resulted in hepatic injury in rats.

Contraindications/Precautions

No information as yet.

Drug-Herb Interactions

No information as yet.

Uses as Food

The leaves are used to make tea or as a culinary spice.

Growth Conditions

Full sun with direct sunlight for 6–8 hours per day. Moderate water. Well-draining soil.

Harvesting

Curry leaf tree may take around 2 years to become established from seed. Once established, the leaves can be harvested frequently and repeatedly.

Ocimum basilicum **L.** (Lamiaceae)

Basil, Garden Basil
罗勒 (Luo Le)

Origin

Originates from tropical Asia, Africa and America

Phytoconstituents

Linalool, methyl chavicol (estragole), methyl cinnamate, methyl eugenol, eugenol, 1,8-cineole, citral, thymol, geraniol, β-caryophyllene, β-elemene, rosmarinic acid and others

Medicinal Uses

<Plant>: Cough, fever, headache, nausea, vomiting, fungal infection, renal disorders, amenorrhoea, insect bites and rheumatism. Used topically for eczema, wounds, ulcers and skin disorders

<Roots>: Influenza, cold, nausea, abdominal pain, migraine, insomnia, indigestion and fatigue. Applied topically for acne, skin infection, anosmia, insect and snake bites

<Leaves>: Fungal infections, delayed menstruation, earache (as ear drops), cough and ringworms

<Seeds>: Haemorrhoids, diarrhoea, dysentery, constipation, ulcers, intestinal and kidney disorders, blurred vision and alleviating pain after childbirth

Pharmacological Activities

Anti-bacterial, anti-cancer, anti-cholinesterase, anti-depressant, anti-diabetic, anti-fungal, anti-inflammatory, anti-obesity, anti-oxidant, anxiolytic, immunomodulatory, nephroprotective and neuroprotective

Clinical Trials

<Basil essential oil applied topically>:
- Higher doses of basil essential oil used over time significantly reduced the pain intensity and frequency of migraine attacks in chronic migraine.

Dosage

No information as yet.

Adverse Reactions

Nausea reported in patients with non-alcoholic fatty liver disease taking *O. basilicum* seeds.

Toxicity

O. basilicum largely demonstrates a good safety profile and is regarded by the Food and Drug Administration as food ("generally regarded as safe"). The main safety concern of *O. basilicum* stems from oral consumption of the essential oil due to carcinogenicity of estragole and methyl eugenol, the major essential oil components.

<Oral administration of 2000 mg/kg estragole>: Reduced water and feed intake, with one death observed in mice.

<*O. basilicum* aerial parts and leaf hydro-alcoholic extracts>: Exerted cytotoxic effect on A375 melanoma cells.

Contraindications/Precautions

The herb should not be used in large amounts in pregnancy, lactation, in children, or for a prolonged period of time due to high amounts of estragole in the essential oil, which have potential mutagenic effects.

Linalool is a potent antibacterial agent, and it is suggested that prolonged use of products containing sub-inhibitory concentrations of essential oil may result in resistant human pathogens and cross protection against antibiotics.

Drug-Herb Interactions

<Methanolic and aqueous extracts of the leaves>: Inhibition of CYP2B6, CYP3A4, CYP3A5, CYP3A7 and CYP2D6 enzymes.

Uses as Food

O. basilicum is widely used as a spice in cuisines around the world and is popular in various cuisines. Basil leaves are used in soups, salads, sauces, and added to bread, fruits, desserts and cocktails.

Growth Conditions

Full sun with direct sunlight for 6–8 hours per day. Moderate water. Well-draining but moisture retentive soil.

Harvesting

Sweet basil leaves should be ready for harvest 6–8 weeks after sowing.

Authors' Notes

There are many varieties of basil, of which one of the most commonly grown and used in Western cuisine is the sweet basil (*O. basilicum*). Thai basil is cultivated in Southeast Asia and is a variety of *O. basilicum*, specifically *O. basilicum* var. *thyrsiflora*. Thai basil has purple stems, light purple flowers with smaller leaves and a licorice-like flavour while sweet basil has green stems, white flowers with broader, delicate leaves and sweet minty aroma.

Ocimum tenuiflorum **L.** (Lamiaceae)

Holy Basil, Sacred Basil, Tulsi / Tulasi
圣罗勒 (Sheng Luo Le)

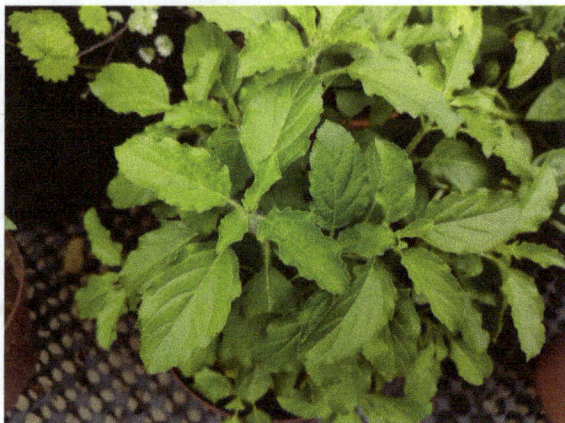

Origin

Originates in tropical Asia, native to India

Phytoconstituents

Methyl chavicol (estragole), methyl eugenol, eugenol, apigenin, α-cubebene, α-copaene, caryophyllene, rosmarinic acid and others

Medicinal Uses

<Plant>: Cough, cold, asthma, bronchitis, digestive disorders, skin problems, eye and ear infections, snake and scorpion bites, arthritis, heart disease, rheumatism, urinary tract infections, headaches and inflammation

<Dried leaves>: According to the Ayurvedic Pharmacopoeia of India, they are used for common cold, cough, hiccups, dyspnea, tastelessness, phthisis, epilepsy, splenic diseases, skin diseases, wounds, helminthiasis, intercostal neuralgia and pleurodynia, abdominal lump and cachexia

Pharmacological Activities

Anti-bacterial, anti-cancer, anti-dementia, anti-diabetic, anti-hyperlipidaemic, anti-inflammatory, anti-microbial, anti-oxidant, anti-proliferative, anti-stress, anti-tuberculosis, enhance cognitive ability and neuroprotective

Clinical Trials

<Tulsi extract capsule>:
- Significant improvement in body weight, body mass index (BMI), lipids, insulin and insulin resistance in obesity.
- Immune modulatory effect.

<Tulsi extract mouthwash>:
- Significantly reduced plaque, gingivitis and bleeding.
- Increased salivary pH and anti-microbial effect on *S. mutans*.

<*O. sanctum* ethanolic leaf extract>:
- Significantly improved cognitive abilities.

<*O. sanctum* toothpaste>:
- Significant reduction in plaque and gingivitis.

<*O. sanctum* gel>:
- Effectively reduced gingival bleeding and inflammation.

Dosage

<1–3 g powdered dried leaves>: Refer to medicinal uses in Ayurvedic Pharmacopoeia of India.

Adverse Reactions

Anti-fertility effects have been reported in male rabbits and female rats.

Toxicity

Various toxicity studies have indicated that *O. tenuiflorum* extracts are safe and non-toxic.

Contraindications/Precautions

Contraindicated in pregnant or lactating women due to conflicting information available on the embryotoxicity of *O. tenuiflorum* leaves. Precaution for children taking *O. tenuiflorum* leaves without medical supervision due to lack of information.

Drug-Herb Interactions

<Eugenol with buthionine sulfoximine>: Hepatotoxicity in mice.

<*O. sanctum* hydroalcoholic extract with valproate>: Enhanced neurobehavioral function and partially reduced oxidative stress.

<*O. sanctum* hydroalcoholic extract with carbamazepine or phenytoin>: Enhanced seizure protection.

Uses as Food

It is a popular herb in Southeast Asia, used fresh or dried. The leaves are used as an aromatic together with garlic, fish sauce and chillies.

Growth Conditions

Full sun with direct sunlight for 6–8 hours per day. Moderate water. Moist soil, well-draining soil.

Harvesting

Harvest of individual leaves or whole stem may begin when the plant has reached 20 cm in height.

Orthosiphon aristatus **(Blume) Miq.** (Lamiaceae)

Cat's Whiskers, Misai Kucing
猫须草 (Mao Xu Cao)

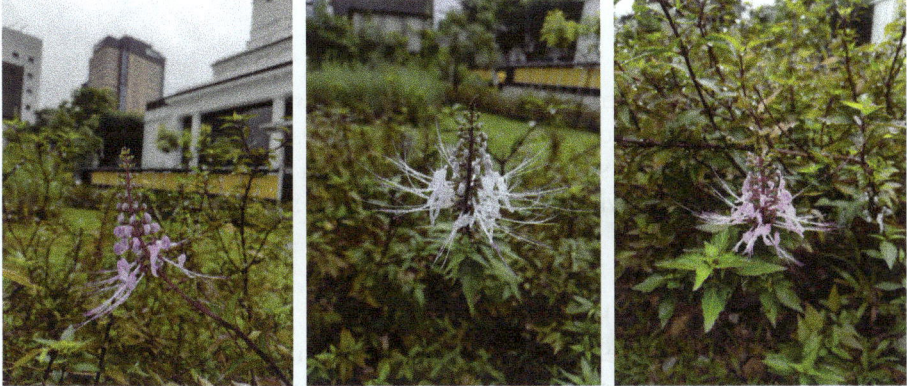

Origin

Originates from tropical Asia

Phytoconstituents

Eupatorin, sinensetin, rosmarinic acid, ursolic acid, caffeic acid derivatives, and others

Medicinal Uses

<Aerial parts>: Oedema, eruptive fevers, gallstones, influenza, jaundice, rheumatism, hepatitis, diabetes, muscle pain and urinary stones

<Leaves>: Diuretic and used to treat hypertension, gout and various urinary disoders such as phosphaturia, haematuria, albuminuria, nephrosis and urinary tract infection

Pharmacological Activities

Analgesic, anti-bacterial, anti-cancer, anti-diabetic, anti-hyperuridaemic, anti-inflammatory, anti-lithic, anti-oxidant, anti-proliferative, diuretic, gastroprotective, neuroprotective and vasorelaxant

Clinical Trials

<Oil in water cosmetic emulsion containing *O. aristatus* leaf extract>:
- Visibly reduce the oily appearance of skin as well as the size of pores, thus leading to a significant improvement of complexion evenness and radiance for individuals with enlarged pores and oily skin.

Dosage

<6–12 g of the dried leaves consumed daily as tea in divided doses>: Promote urination as adjuvant therapy in minor urinary tract disorders in adults and elderly.

<6–12 g of the dried leaves and stem tips>: Irrigation therapy for inflammatory and bacterial diseases of the lower urinary tract and kidney stones. Ensure sufficient fluid intake of at least 2 L daily during treatment.

Adverse Reactions

A clinical trial found that consumption of the dried plant as tea was well-tolerated.

Toxicity

Toxicity studies indicate that various extracts of *O. aristatus* in rats were found to be safe and non-toxic.

Contraindications/Precautions

The plant contains high amounts of potassium and is not recommended for people with heart disease. It should not be used as irrigation therapy in oedema due to impaired heart and renal function. Use under 18 years of age and in pregnancy and lactation is not recommended due to lack of data. Consult a doctor or healthcare professional if symptoms including fever, dysuria, spasms, or blood in the urine, occur during the use of the medicinal product.

Drug-Herb Interactions

<*O. aristatus* extracts>: Inhibition of human UGT isoforms, CYP2C19, CYP2D6 and CYP3A4 *in vitro*.

<Nuvastatic™ with gemcitabine>: Significantly reduced pancreatic tumour growth in mice.

Uses as Food

Although the leaves are edible, *O. aristatus* is not normally consumed as food, apart from consumption as tea. Java tea is made with the leaves of Cat's Whiskers and is commercially available.

Growth Conditions

Full sun with direct sunlight for 6–8 hours per day. Moderate water. Well-draining soil, high organic content soil.

Harvesting

Both the leaves and stem can be harvested 8–10 weeks after planting, at the start of flowering.

Authors' Notes

The plant produces attractive white or purple flowers that resemble whiskers of cats, hence, its common name is Cat's Whiskers.

Pandanus amaryllifolius **Roxb.** (Pandanaceae)

Pandan, Fragrant Pandan
香兰叶 (Xiang Lan Ye)

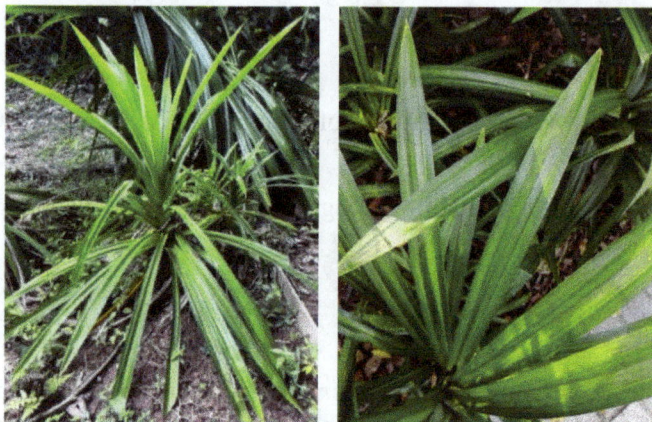

Origin

Originates from Southeast Asia

Phytoconstituents

2-acetyl-1-pyrroline, 6E-pandanamine, pandanin, pandamarilactonines, pandamarilactones, 3-methyl-2-(5H)-furanone, gallic acid, catechin, kaempferol, etc.

Medicinal Uses

Gonorrhoea, fever, syphilis, anaemia, dengue, indigestion and flatulence

Pharmacological Activities

Anti-bacterial, anti-cancer, ant-idiabetic, anti-oxidant, anti-viral, anti-amyloidogenic, improve metabolic syndrome and neuroprotective

Clinical Trials

<*P. amaryllifolius* tea containing dried leaf powder in water>:
- Significantly lower mean postprandial plasma glucose concentration peak.

Dosage

No information as yet.

Adverse Reactions

No information as yet.

Toxicity

No information as yet.

Contraindications/Precautions

No information as yet.

Drug-Herb Interactions

No information as yet.

Uses as Food

Pandan leaves are not consumed directly. In Southeast Asia, the leaves are commonly cooked together with rice or desserts for its aroma, or made into tea. Pandan is also used in bakery products and sweets for its flavour. Pandan extract may be used in cakes, kuehs, agar-agar, waffles and other pastries.

Growth Conditions

Full sun with direct sunlight for 6–8 hours per day. Moderate water. Well-draining soil, fertile loamy soil.

Harvesting

Harvest of Pandan leaves may begin around 6 months after planting.

Pereskia bleo (Kunth) DC. (Cactaceae)

Rose Cactus, Wax Rose, Seven Star Needle
七星针 (Qi Xing Zhen)

Origin

Native to Panama, Central and Western South America

Phytoconstituents

Vitexin, bleogen pB1, β-sitosterol, carotenoids, catechin, myricetin, quercetin and others

Medicinal Uses

Hypertension, diabetes, cancer, rheumatism, headache, haemorrhoids, gastric pain and asthma

Pharmacological Activities

Analgesic, anti-bacterial, anti-cancer, anti-fungal, anti-oxidant, anti-proliferative, mosquito larvicidal, vasorelaxant and wound healing

Clinical Trials

No information as yet.

Dosage

No information as yet.

Adverse Reactions

No information as yet.

Toxicity

<Aqueous leaf fraction of 1–165 µg/mL>: Mutagenic metabolites may be generated when the aqueous leaf fraction is metabolised by liver enzymes.

Contraindications/Precautions

No information as yet.

Drug-Herb Interactions

No information as yet.

Uses as Food

Leaves are eaten raw as a vegetable or salad. The plant is also sometimes used as a spice.

Growth Conditions

Full sun with direct sunlight for 6–8 hours per day. Moderate water. Well-draining soil.

Harvesting

A stem cutting can take around 7 months to start flowering. Both the leaves and the mature, yellow, funnel-shaped fruits of Wax Rose can be harvested for use.

Authors' Notes

Pereskia bleo is also known as *Leuenbergeria bleo*. It is a leafy cactus. Its Chinese name Qi Xing Zhen literally means "Seven star needle", because of its thorny stems. The pricks can be very painful, so please be careful.

Plectranthus amboinicus (Lour.) Spreng.
(Lamiaceae)

Indian Borage, Cuban Oregano
到手香 (Dao Shou Xiang)

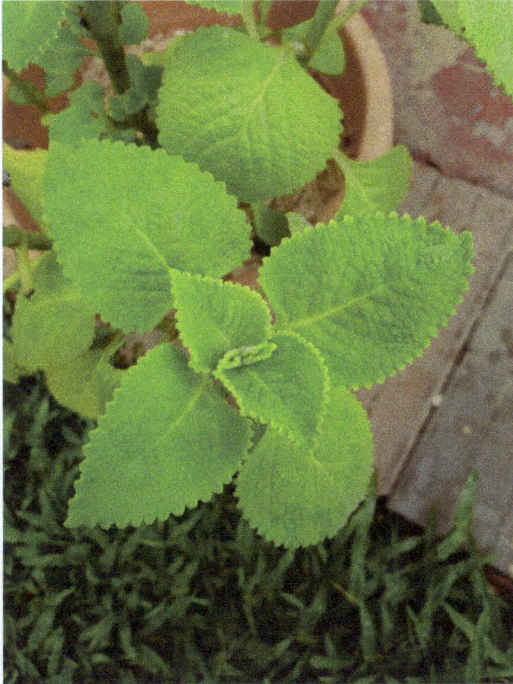

Origin

The plant is of unknown origin but possibly from Africa and India

Phytoconstituents

Thymol, carvacrol, caryophyllene, cymene, apigenin, cirsimaritin, salvigenin, quercetin, rosmarinic acid and others

Medicinal Uses

Headaches, hyperthermia, coryza, influenza, laryngitis, hoarseness, bronchitis, asthma, haemoptysis, epistaxis, haematemesis, gastralgia, flatulence, colic, dyspepsia, angina and convulsions

Pharmacological Activities

Analgesic, anti-arthritic, anti-bacterial, anti-cancer, anti-diabetic, anti-fungal, anti-inflammatory, anti-malarial, anti-oxidant, insecticidal and mosquito repellent

Clinical Trials

No information as yet.

Dosage

No information as yet.

Adverse Reactions

There have been reports of allergic contact dermatitis.

Toxicity

A toxicity study indicated that acute doses of up to 5000 mg/kg 95% ethanolic leaf extract in rats resulted in no deaths and adverse effects.

Contraindications/Precautions

No information as yet.

Drug-Herb Interactions

No information as yet.

Uses as Food

In India, the plant is commonly used for culinary purposes. The leaves can also be eaten raw.

Growth Conditions

Semi-shade with direct sunlight for 4–6 hours per day. Moderate water. Well-draining soil, sandy soil.

Harvesting

Indian Borage has light-green, simple leaves that are thick and fleshy. Harvest the leaves for use when desired.

Punica granatum L. (Lythraceae)

Pomegranate, Granada
石榴 (Shi Liu)

Origin

Native to Iran and the Himalayas in North India

Phytoconstituents

Ellagitannin, punicalagin, punicalin, anthocyanins, flavanols, ellagic acid, punicic acid and others

Medicinal Uses

<Fruit rind>: According to the Chinese Pharmacopoeia, it is used to treat chronic diarrhoea and dysentery, blood stool, prolapse of the rectum, flooding and spotting, vaginal discharge, and abdominal pain caused by worm accumulation

<Fruit juice>: Fever, dyspepsia, leprosy and cataracts

Pharmacological Activities

Anti-arthritic, anti-bacterial, anti-diabetic, anti-fungal, anti-hyperlipidaemic, anti-hypertensive, anti-inflammatory, anti-obesity, anti-oxidant, anti-proliferative, antiviral, cardioprotective, chemoprotective and neuroprotective

Clinical Trials

<Pomegranate juice>:
- Significant decrease in mean systolic blood pressure in metabolic syndrome, postprandial glycaemic response, diastolic blood pressure, LDL cholesterol and improved lipid metabolism in dyslipidaemia.
- Cardioprotective effect in unstable angina
- Anti-oxidant activity in type 2 diabetes

<Pomegranate extract capsules>:
- Significant decrease in weight, fasting serum glucose, insulin, LDL cholesterol and triglycerides in obesity, inflammatory factors and oxidative stress biomarkers, total anti-oxidant capacity increased and lipid profile improved significantly in type 2 diabetes.

<Pomegranate seed powder tea bags>:
- Significantly decreased fasting blood glucose and HbA1c in type 2 diabetes.

<Pomegranate supplement>:
- Improved menopausal symptoms and quality of life

Dosage

<3–9 g of fruit rind decocted with water taken orally>: Refer to medicinal uses in Chinese Pharmacopoeia.

Adverse Reactions

No information as yet.

Toxicity

Various toxicity studies indicated that *P. granatum* extracts and juice were safe and non-toxic in murine models except for the following:

<Intraperitoneal dosing of dried methanolic extract of *P. granatum* rind dissolved in water at 1000 mg/kg>: Lethal to mice.

Contraindications/Precautions

Precautions with fruit rind and dried pericarp in high doses as they are slightly toxic.

Drug-Herb Interactions

<Pomegranate juice with metformin>: Significantly reduced metformin plasma concentration, potentially reducing the efficacy of metformin.

<Pomegranate juice with rifampin or isoniazid>: Synergistic antibacterial activity against multi-drug resistant *M. tuberculosis* clinical isolates.

<Pomegranate peel extract with aspirin>: Synergistic anti-nociceptive effect *in vivo* in rats.

Uses as Food

The fruit is consumed fresh or processed into syrup, sauces, beverages and desserts. The seeds can also be used as a spice, or a garnish.

Growth Conditions

Full sun with direct sunlight for 6–8 hours per day. Moderate water. Well-draining soil, fertile loamy soil.

Harvesting

Plant that is grown from seed should start fruiting when it is 3 years old and fruits should ripen in 3–6 months after flowers appear. Pomegranate produces red flowers and ripe fruits are red-brown in colour.

Rosmarinus officinalis **L.** (Lamiaceae)

Rosemary, Romero
迷迭香 (Mi Die Xiang)

Origin

Originates in Europe and along the African coast of the Mediterranean Sea

Phytoconstituents

Carnosic acid, carnosol, camphor, chlorogenic acid, oleanolic acid, rosmarinic acid, ursolic acid, 1,8-cineole, α-pinene, β-pinene, apigenin and others

Medicinal Uses

<Plant>: Anti-spasmodic in renal colic and dysmenorrhoea, carminative, expectorant, hair growth stimulant and treatment of cognitive disorders

<Essential oil and leaves>: Cholagogue, diaphoretic, digestant, diuretic, emmenagogue, laxative and tonic. Used for headache, menstrual disorders, fatigue, memory disorders, sprains and bruises

Pharmacological Activities

Anti-ageing, anti-arthritic, anti-bacterial, anti-cancer, anti-depressant, anti-diabetic, anti-inflammatory, anti-oxidant, anxiolytic, cardioprotective, hepatoprotective, nephroprotective and neuroprotective

Clinical Trials

<Rosemary essential oil via inhalation>:
- Significantly increased focus, concentration and alertness.
- Decreased sleepiness.

<Capsules with dried powdered aerial parts of rosemary>:
- Significantly reduced anxiety and depression, boosted prospective and retrospective memory and improved sleep quality.

<Rosemary essential oil on a sugar cube>:
- Significantly increased blood pressure in hypotension.

Dosage

<Infusion of 2 g in 150 mL water three times daily>: Dyspeptic complaints.

<Bath additive by decocting 50 g of leaf in 1 L water or ointment (6–10% essential oil in base of petroleum jelly or lanolin)>: Supportive therapy for rheumatic diseases and circulatory problems.

Adverse Reactions

Dried leaves have been observed to cause contact allergy.

Toxicity

Generally, toxicity studies have indicated that extracts of *R. officinalis* were safe and non-toxic except for the following:

<2400 mg/kg *R. officinalis* supercritical CO_2 extract and 3800 mg/kg acetone extract for 90 days>: Reversible liver enlargement in rats was observed.

Contraindications/Precautions

Not recommended during pregnancy. Rosemary oil preparations are contraindicated in obstruction of bile duct, cholangitis, liver disease, gallstones and other biliary disorders.

Drug-Herb Interactions

<*R. officinalis* supercritical CO_2 extract>: Reversible induction of CYP2A1, CYP2A2, CYP2C11, CYP2E1 and CYP4A enzymes.

<*R. officinalis* leaf extract with clindamycin>: Synergistic antibacterial effect against 2 methicillin-resistant *Staphylococcus aureus* strains *in vitro*.

<Rosemary essential oil with diclofenac sodium topical gel>: Enhanced analgesic effect.

<Carnosol>: Inhibited CYP1A2, CYP2C9, CYP2C19, CYP2D6 and CYP3A4 enzymes expressed in baculosomes and modulated metabolic enzymes and transporters *in vitro*.

<Carnosic acid with carmustine, lomustine and cisplatin>: Enhanced anticancer effects against melanoma *in vitro* and *in vivo*.

Uses as Food

Rosemary is a spice widely used to flavour dishes.

Growth Conditions

Full sun with direct sunlight for 6–8 hours per day. Little water. Well-draining soil, lightweight soil, dry soil and sandy soil.

Harvesting

While leaves can be harvested regularly from established plants, leaves harvested shortly after flowers appear will be most flavourful and aromatic.

Strobilanthes crispa **(L.) Blume** (Acanthaceae)

Black Face General, Bayam Karang
黑面将军 (Hei Mian Jiang Jun)

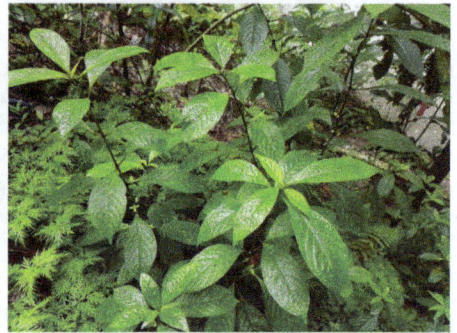

Origin

Native to the regions from Madagascar to the Malay Archipelago

Phytoconstituents

Phenolic acids, flavonoids, stigmasterol, β-sitosterol, saponins, and others

Medicinal Uses

<Leaves and stems>: General health promotion, detoxification, to lower blood pressure, boost immunity and prevent cancer

Pharmacological Activities

Anti-cancer, anti-diabetic, anti-glycolytic, anti-hyperlipidaemic, anti-microbial, anti-oxidant, anti-proliferative, anti-tumour, chemopreventive, immunomodulatory and wound healing

Clinical Trials

No information as yet.

Dosage

No information as yet.

Adverse Reactions

Some individuals experienced side effects, which included a "needle poking" sensation at cancer-affected areas, some swelling and pain around the knees, ankles or legs, which persisted for 1–2 weeks.

Toxicity

Toxicity studies have indicated that *S. crispa* leaf extracts were safe and not cytotoxic.

Contraindications/Precautions

No information as yet.

Drug-Herb Interactions

<*S. crispa* bioactive sub-fraction>: Low inhibition of CYP2B6, CYP2C19, CYP2D6 and CYP3A4 enzymes.

Uses as Food

The leaves of *S. crispa* are rough and are usually consumed as a drink (beverage) by boiling the leaves.

Growth Conditions

Full sun with direct sunlight for 6–8 hours per day. Moderate water. Well-draining soil, moist soil.

Harvesting

As the leaves are used, harvest when the foliage is healthy. The foliage should appear dark green and shiny.

Vernonia amygdalina **Delile** (Asteraceae)

Bitter Leaf, South African Leaf
南非叶 (Nan Fei Ye)

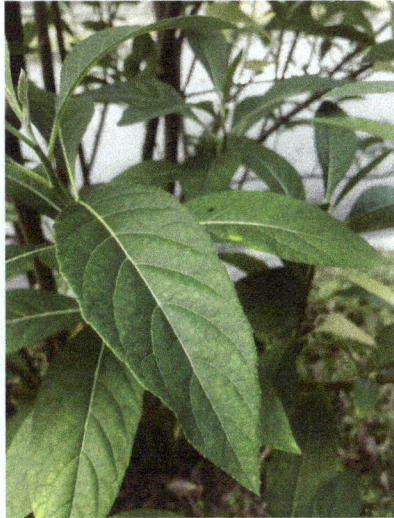

Origin

Native to tropical Africa

Phytoconstituents

Hydroxyvernolide, vernodalin, vernodalol, vernolide, veronicoside, saponins and others

Medicinal Uses

<Roots>: Impotence and to induce labour

<Plant>: Malaria, fever, cough, diarrhoea, dysentery, rheumatism, rashes, wounds, headache, hepatitis, diabetes, digestive disorders, piles, hypertension, nausea, and as an anthelmintic, laxative and fertility inducer

Pharmacological Activities

Anti-bacterial, anti-cancer, anti-diabetic, anti-inflammatory, anti-malarial, anti-obesity, anti-oxidant, gastroprotective, hepatoprotective, immunomodulatory, nephroprotective and neuroprotective

Clinical Trials

<Immunace® capsules and/or *V. amygdalina* aqueous leaf extract>:
- Significant improvement in skin rashes and immunological effect in HIV.

Dosage

No information as yet.

Adverse Reactions

Nocturia, insomnia and cough were observed following consumption of a leaf infusion.

Toxicity

<Fresh leaf infusion>: Anaemia in some patients with malara; *in vitro* haemolytic activity on human erythrocytes, especially in sickle-cell subjects with the genotype Hb-SS.

<Oral administration of 500–2000 mg/kg aqueous leaf extract for 14 days>: Significantly reduced erythrocyte count in rats.

Contraindications/Precautions

No information as yet.

Drug-Herb Interactions

<Aqueous leaf extract and digoxin>: Increased intestinal digoxin absorption by inhibition of P-glycoprotein.

<Methanol and aqueous leaf extracts>: Inhibition of major human cytochrome P450 enzymes, including CYP3A4, CYP1A2, CYP2C8, CYP2C9 and CYP2D6.

Uses as Food

In many parts of Africa, leaves are used to prepare soup, such as bitter leaf soup, or consumed as a vegetable. Fresh leaves are also used to make bitter leaf juice.

Growth Conditions

Full sun with direct sunlight for 6–8 hours per day. Moderate water. Well-draining humus-rich soil, dry soil.

Harvesting

A rooted cutting can take 2 months to grow to 25 cm tall, with 2 shoots.

Vitex trifolia **L.** (Lamiaceae)

Simpleleaf Chastetree, Common Blue Vitex
三叶蔓荆 (San Ye Man Jing), 蔓荆子 (Man Jing Zi)

Origin

Originated in Southeast Asia and the Pacific islands

Phytoconstituents

Artemetin, casticin, vitexilactone, vitetrifolin, maslinic acid and others

Medicinal Uses

<Leaves>: Intestinal disorders

<Dried ripe fruits>: According to the Chinese Pharmacopoeia, they are used for headache, common cold, painful swelling of gums, inflammation of the eye with lacrimation, blurring of vision, dizziness, upper respiratory infections, rhinitis, conjunctivitis and otitis media

Pharmacological Activities

Anagelsic, anti-asthmatic, anti-bacterial, anti-cancer, anti-inflammatory, anti-neoplastic, anti-oxidant, anti-proliferative, anti-tumour, hepatoprotective, inhibit acetylcholinesterase and insulin sensitising

Clinical Trials

No information as yet.

Dosage

<5–9 g dried mature fruits decocted with water or wine, taken orally>: Refer to medicinal uses in Chinese Pharmacopoeia.

Adverse Reactions

No information as yet.

Toxicity

No information as yet.

Contraindications/Precautions

No information as yet.

Drug-Herb Interactions

No information as yet.

Uses as Food

The leaves are used to prepare decoction and tea, which are consumed for health promotion.

Growth Conditions

Full sun with direct sunlight for 6–8 hours per day. Moderate water. Moist soil, well-draining soil.

Harvesting

Leaves, fruits, flowers and roots can be harvested when required once the plant is established.

Appendix

List of Plants by Scientific Name

Scientific Name	Common Name
Abelmoschus esculentus	Lady's Finger, Okra, 羊角豆 (Yang Jiao Dou)
Aloe vera	Aloe, Lidah Buaya, 芦荟 (Lu Hui)
Alternanthera sessilis	Sessile Joyweed, 红田乌草 (Hong Tian Wu Cao)
Andrographis paniculata	Bitterweed, King of Bitters, Common Andrographis 穿心莲 (Chuan Xin Lian), 苦草 (Ku Cao)
Azadirachta indica	Neem Tree, Indian Lilac, Margosa Tree, 印度苦楝 (Yin Du Ku Lian)
Carica papaya	Papaya, Papaw, 木瓜 (Mu Gua)
Centella asiatica	Asiatic Pennywort, Indian Pennywort, Gotu Kola, 积雪草 (Ji Xue Cao)
Clitoria ternatea	Butterfly Pea, Blue Pea, 蝴蝶花豆 (Hu Die Hua Dou)
Cymbopogon citratus	Lemongrass, Citronella Grass, 香茅 (Xiang Mao)
Gynura procumbens	Longevity Spinach, Sambung Nyawa, 尖尾凤 (Jian Wei Feng), 蛇接骨 (She Jie Gu)
Hibiscus sabdariffa	Roselle, Sour Tea, Jamaica Sorrel, 洛神葵 (Luo Shen Kui)
Impatiens balsamina	Balsam, 凤仙花 (Feng Xian Hua)
Ipomoea batatas	Sweet Potato, Keledek, 甘薯 (Gan Shu)
Mentha spicata	Spearmint, Common Mint, Bo He, 留兰香 (Liu Lan Xiang)
Momordica charantia	Bitter Gourd, Bitter Melon, 苦瓜 (Ku Gua)
Morinda citrifolia	Noni, Cheese Fruit, 海巴戟 (Hai Ba Ji)
Moringa oleifera	Horseradish Tree, Drumstick Tree, 辣木 (La Mu)
Morus alba	White Mulberry, Tut, 桑 (Sang)
Murraya koenigii	Curry Leaf Tree, Indian Curry Tree, 麻绞叶 (Ma Jiao Ye), 咖喱叶 (Ga Li Ye)

Scientific Name	Common Name
Ocimum basilicum	Basil, Garden Basil, 罗勒 (Luo Le)
Ocimum tenuiflorum	Holy Basil, Sacred Basil, Tulsi/Tulasi, 圣罗勒 (Sheng Luo Le)
Orthosiphon aristatus	Cat's Whiskers, Misai Kucing, 猫须草 (Mao Xu Cao)
Pandanus amaryllifolius	Pandan, Fragrant Pandan, 香兰叶 (Xiang Lan Ye)
Pereskia bleo	Rose Cactus, Wax Rose, Seven Star Needle, 七星针 (Qi Xing Zhen)
Plectranthus amboinicus	Indian Borage, Cuban Oregano, 到手香 (Dao Shou Xiang)
Punica granatum	Pomegranate, Granada, 石榴 (Shi Liu)
Rosmarinus officinalis	Rosemary, Romero, 迷迭香 (Mi Die Xiang)
Strobilanthes crispa	Black Face General, Bayam Karang, 黑面将军 (Hei Mian Jiang Jun)
Vernonia amygdalina	Bitter Leaf, South African Leaf, 南非叶 (Nan Fei Ye)
Vitex trifolia	Simpleleaf Chastetree, Common Blue Vitex, 三叶蔓荆 (San Ye Man Jing), 蔓荆子 (Man Jing Zi)

List of Plants by Common Name

Common Name	Scientific Name
Aloe	*Aloe vera*
Asiatic Pennywort	*Centella asiatica*
Balsam	*Impatiens balsamina*
Basil	*Ocimum basilicum*
Bitterweed	*Andrographis paniculata*
Bitter Gourd	*Momordica charantia*
Bitter Leaf	*Vernonia amygdalina*
Black Face General	*Strobilanthes crispa*
Butterfly Pea	*Clitoria ternatea*
Cat's Whiskers	*Orthosiphon aristatus*
Curry Leaf Tree	*Murraya koenigii*
Holy Basil	*Ocimum tenuiflorum*
Horseradish Tree	*Moringa oleifera*
Indian Borage	*Plectranthus amboinicus*
Lady's Finger	*Abelmoschus esculentus*
Lemongrass	*Cymbopogon citratus*
Longevity Spinach	*Gynura procumbens*
Neem Tree	*Azadirachta indica*
Noni	*Morinda citrifolia*
Pandan	*Pandanus amaryllifolius*
Papaya	*Carica papaya*
Pomegranate	*Punica granatum*
Rose Cactus	*Pereskia bleo*
Roselle	*Hibiscus sabdariffa*
Rosemary	*Rosmarinus officinalis*

Common Name	Scientific Name
Sessile Joyweed	*Alternanthera sessilis*
Simpleleaf Chastetree	*Vitex trifolia*
Spearmint	*Mentha spicata*
Sweet Potato	*Ipomoea batatas*
White Mulberry	*Morus alba*

References

Abelmoschus esculentus

Doreddula, S. K., Bonam, S. R., Gaddam, D. P., *et al.* (2014) Phytochemical analysis, anti-oxidant, anti-stress, and nootropic activities of aqueous and methanolic seed extracts of ladies finger (*Abelmoschus esculentus* L.) in mice. *Sci. World J.*, 519848.

Gou, L., Liu, G., Ma, R., *et al.* (2020) High fat-induced inflammation in vascular endothelium can be improved by *Abelmoschus esculentus* and metformin via increasing the expressions of miR-146a and miR-155. *Nutr. Metab.*, *17*, 35.

Moradi, A., Tarrahi, M. J., Ghasempour, S., *et al.* (2020) The effect of okra (*Abelmoschus esculentus*) on lipid profiles and glycemic indices in type 2 diabetic adults: Randomized double blinded trials. *Phytother. Res.*, *34(12)*, 3325–3332.

Aloe vera

Gao, Y., Kuok, K. I., Jin, Y., Wang, R. (2019) Biomedical applications of *Aloe vera*. *Crit. Rev. Food Sci. Nutr.*, *59(sup1)*, S224–S256.

Hu, Y. H., Quan, Z. Y., Li, D. K., *et al.* (2022) Inhibition of CYP3A4 enhances aloe-emodin induced hepatocyte injury. *Toxicol. In Vitro*, *79*, 105–276.

Kurian, I. G., Dileep, P., Ipshita, S., Pradeep, A. R. (2018) Comparative evaluation of subgingivally delivered 1% metformin and *Aloe vera* gel in the treatment of intrabony defects in chronic periodontitis patients: A randomized, controlled clinical trial. *J. Investig. Clin. Dent.*, *9(3)*, e12324.

Liu, Y., Mapa, M. S. T., Sprando, R. L. (2021) Anthraquinones inhibit cytochromes P450 enzyme activity *in silico* and *in vitro*. *J. Appl. Toxicol.*, *41(9)*, 1438–1445.

Alternanthera sessilis

Hwong, C. S., Leong, K. H., Abdul Aziz, A., *et al.* (2022) Alternanthera sessilis: Uncovering the nutritional and medicinal values of an edible weed. *J. Ethnopharmacol.*, *298*, 115608.

Kota, S., Govada, V. R., Anantha, R. K., Verma, M. K. (2017) An investigation into phytochemical constituents, anti-oxidant, anti-bacterial and anti-cataract activity of *Alternanthera sessilis*, a predominant wild leafy vegetable of South India. *Biocatal. Agric. Biotechnol.*, *10*, 197–203.

Muniandy, K., Gothai, S., Badran, K. M. H., *et al.* (2018) Suppression of pro-inflammatory cytokines and mediators in LPS-Induced RAW 264.7 macrophages by stem extract of *Alternanthera sessilis* via the inhibition of the NF-κB Pathway. *J. Immunol. Res.*, 3430684.

Andrographis paniculata

Hossain, M. S., Urbi, Z., Sule, A., Hafizur Rahman, K. M. (2014) *Andrographis paniculata* (Burm. f.) Wall. ex Nees: A review of ethnobotany, phytochemistry, and pharmacology. *Sci. World J.*, *2014*, 274905.

Hu, X. Y., Wu, R. H., Logue, M., *et al.* (2017) *Andrographis paniculata* (chuān xīn lian) for symptomatic relief of acute respiratory tract infections in adults and children: A systematic review and meta-analysis. *PLoS One*, *12(8)*, e0181780.

Okhuarobo, A., Falodun, J. E., Erharuyi, O., *et al.* (2014) Harnessing the medicinal properties of *Andrographis paniculata* for diseases and beyond: A review of its phytochemistry and pharmacology. *Asian Pac. J. Trop. Dis.*, *4(3)*, 213–222.

Saxena, R. C., Singh, R., Kumar, P., *et al.* (2010) A randomized double blind placebo controlled clinical evaluation of extract of *Andrographis paniculata* (KalmCold) in patients with uncomplicated upper respiratory tract infection. *Phytomedicine*, *17(3–4)*, 178–185.

Shang, Y. X., Shen, C., Stub, T., *et al.* (2022) Adverse effects of andrographolide derivative medications compared to the safe use of herbal preparations of *Andrographis paniculata*: Results of a systematic review and meta-analysis of clinical studies. *Front. Pharmacol.*, *13*, 773282.

Azadirachta indica

Amaeze, O., Eng, H., Horlbogen, L., *et al.* (2021) Cytochrome P450 enzyme inhibition and herb-drug interaction potential of medicinal plant extracts used for management of diabetes in Nigeria. *Eur. J. Drug Metab. Pharmacokinet.*, *46(3)*, 437–450.

Ashafa, A. O., Orekoya, L. O., Yakubu, M. T. (2012) Toxicity profile of ethanolic extract of *Azadirachta indica* stem bark in male Wistar rats. *Asian Pac. J. Trop. Biomed.*, *2(10)*, 811–817.

Irais, C. M., Claudia, B. R., David, P. E., *et al.* (2021) Leaf and fruit methanolic extracts of *Azadirachta indica* exhibit anti-fertility activity on rats' sperm quality and testicular histology. *Curr. Pharm. Biotechnol.*, *22(3)*, 400–407.

Pingali, U., Ali, M. A., Gundagani, S., Nutalapati, C. (2020) Evaluation of the effect of an aqueous extract of *Azadirachta indica* (neem) leaves and twigs on glycemic control, endothelial dysfunction and systemic inflammation in subjects with type 2 diabetes mellitus — A randomized, double-blind, placebo-controlled clinical study. *Diabetes Obes. Metab.*, *13*, 4401–4412.

Carica papaya

Hamed, A. N. E., Abouelela, M. E., El Zowalaty, *et al.* (2022) Chemical constituents from *Carica papaya* Linn. leaves as potential cytotoxic, EGFR^{wt} and aromatase (CYP19A) inhibitors; A study supported by molecular docking. *RSC Adv.*, *12(15)*, 9154–9162.

Kasture, P. N., Nagabhushan, K. H., Kumar, A. (2016) A multi-centric, double-blind, placebo-controlled, randomized, prospective study to evaluate the efficacy and safety of *Carica papaya* leaf extract, as empirical therapy for thrombocytopenia associated with dengue fever. *J. Assoc. Physicians India*, *64(6)*, 15–20.

Lim, X. Y., Chan, J. S. W., Japri, N., *et al.* (2021) *Carica papaya* L. leaf: A systematic scoping review on biological safety and herb-drug interactions. *Evid. Based Complementary Altern. Med.*, *2021*, 5511221.

Natural Medicines (n.d.) *Papaya professional monograph* [updated 17 May 2018]. Available at https:// naturalmedicines.therapeuticresearch. com/databases/food,-herbs-supplements/ professional.aspx?productid=488 (accessed 17 September 2018).

Ojo, O. A., Ojo, A. B., Awoyinka, O., *et al.* (2017) Aqueous extract of *Carica papaya* Linn. roots potentially attenuates arsenic induced biochemical and genotoxic effects in Wistar rats. *J. Tradit. Complement. Med.*, *8(2)*, 324–334.

Centella asiatica

Bandopadhyay, S., Mandal, S., Ghorai, M., *et al.* (2023) Therapeutic properties and pharmacological activities of asiaticoside and madecassoside: A review. *J. Cell. Mol. Med.*, *27(5)*, 593–608.

European Medicines Agency, Committee on Herbal Medicinal Products (2022) *Assessment report on Centella asiatica (L.) Urb., herba — Revision 1*. Available at https://www.ema.europa.eu/en/ documents/herbal-report/assessment-report-centella-asiatica-l-urb-herba-revision-1_en.pdf (accessed 4 January 2023).

Gohil, K. J., Patel, J. A., Gajjar, A. K. (2010) Pharmacological review on

Centella asiatica: A potential herbal cure-all. *Indian J. Pharm. Sci.*, *72(5)*, 546–556.

Kumar, R., Arora, R., Sarangi, S. C., *et al.* (2021) Pharmacodynamic and pharmacokinetic interactions of hydroalcoholic leaf extract of *Centella asiatica* with valproate and phenytoin in experimental models of epilepsy in rats. *J. Ethnopharmacol.*, *270*, 113784.

Clitoria ternatea

Chusak, C., Thilavech, T., Henry, C. J., Adisakwattana, S. (2018) Acute effect of *Clitoria ternatea* flower beverage on glycemic response and anti-oxidant capacity in healthy subjects: A randomized crossover trial. *BMC Complement. Altern. Med.*, *18(1)*, 6.

Taranalli, A. D., Cheeramkuzhy, T. C. (2000) Influence *of Clitoria ternatea* extracts on memory and central cholinergic activity in rats. *Pharm. Biol.*, *38(1)*, 51–56.

Cymbopogon citratus

Hacke, A. C. M., D'Avila da Silva, F., Lima, D., *et al.* (2022) Cytotoxicity of *Cymbopogon citratus* (DC) Stapf fractions, essential oil, citral, and geraniol in human leukocytes and erythrocytes. *J. Ethnopharmacol.*, *291*, 115147.

Nguyen, C., Mehaidli, A., Baskaran, K., *et al.* (2019) Dandelion root and lemongrass extracts induce apoptosis, enhance chemotherapeutic efficacy, and reduce tumour xenograft growth *in vivo* in prostate cancer. *Evid. Based Complementary Altern. Med.*, *2019*, 2951428.

Ruvinov, I., Nguyen, C., Scaria, B., *et al.* (2019) Lemongrass extract possesses potent anticancer activity against human colon cancers, inhibits tumorigenesis, enhances efficacy of FOLFOX, and reduces its adverse effects. *Integr. Cancer Ther.*, *18*, 1534735419889150.

Soonwera, M., Phasomkusolsil, S. (2015) Efficacy of Thai herbal essential oils as green repellent against mosquito vectors. *Acta Trop.*, *142*, 127–130.

Gynura procumbens

Amin, M. Z., Afrin, M., Meghla, N. S., *et al.* (2021) Assessment of anti-bacterial, anti-inflammatory, and cytotoxic effects of different extracts of *Gynura procumbens* leaf. *Curr. Ther. Res. Clin. Exp.*, *95*, 100636.

Husni, Z., Ismail, S., Zulkiffli, M. H., *et al.* (2017) *In vitro* inhibitory effects of *Andrographis paniculata*, *Gynura procumbens*, *Ficus deltoidea*, and *Curcuma xanthorrhiza* extracts and constituents on human liver glucuronidation activity. *Pharmacogn. Mag.*, *13(Suppl 2)*, S236–S243.

Jarikasem, S., Charuwichitratana, S., Siritantikorn, S., *et al.* (2013) Anti-herpetic effects of *Gynura procumbens*. *Evid. Based Complementary Altern. Med.*, *2013*, 394865.

Hibiscus sabdariffa

Al-Anbaki, M., Nogueira, R. C., Cavin, A. L., *et al.* (2019) Treating uncontrolled hypertension with *Hibiscus sabdariffa* when standard treatment is insufficient: Pilot intervention. *J. Tradit. Complement. Med.*, *25(12)*, 1200–1205.

Njinga, N. S., Kola-Mustapha, A. T., Quadri, A. L., *et al.* (2020) Toxicity assessment of sub-acute and sub-chronic oral administration and diuretic potential of aqueous extract of *Hibiscus sabdariffa* calyces. *Heliyon*, *6(9)*, e04853.

Nwachukwu, D. C., Aneke, E. I., Nwachukwu, N. Z., *et al.* (2017) Does consumption of an aqueous extract of *Hibscus sabdariffa* affect renal function in subjects with mild to moderate hypertension?. *J. Physiol. Sci.*, *67, 227–234.

Riaz, G., Chopra, R. (2018) A review on phytochemistry and therapeutic uses of *Hibiscus sabdariffa* L. *Biomed. Pharmacother.*, *102, 575–586.

Impatiens balsamina

Jiang, H. F., Zhuang, Z. H., Hou, B. W., *et al.* (2017) Adverse effects of hydroalcoholic extracts and the major components in the stems of *Impatiens balsamina* L. on *Caenorhabditis elegans*. *Evid. Based Complementary Altern. Med.*, *2017*, 4245830.

Qian, H., Wang, B., Ma, J., *et al.* (2023) *Impatiens balsamina*: An updated review on the ethnobotanical uses, phytochemistry, and pharmacological activity. *J. Ethnopharmacol.*, *303*, 115956.

Yang, X., Guo, T., Du, Z., *et al.* (2023) Protective effects of MNQ against lipopolysaccharide-induced inflammatory damage in bovine ovarian follicular granulosa cells *in vitro*. *J. Steroid Biochem. Mol. Biol.*, *230*, 106274.

Ipomoea batatas

Jiang, T., Zhou, J., Liu, W., *et al.* (2020) The anti-inflammatory potential of protein-bound anthocyanin compounds from purple sweet potato in LPS-induced RAW264.7 macrophages. *Food Res. Int.*, *137*, 109647.

Luis, G., Rubio, C., Gutiérrez, A. J., *et al.* (2014) Evaluation of metals in several varieties of sweet potatoes (*Ipomoea batatas* L.): Comparative study. *Environ. Monit. Assess.*, *186(1)*, 433–440.

Oki, T., Kano, M., Ishikawa, F., *et al.* (2017) Double-blind, placebo-controlled pilot trial of anthocyanin-rich purple sweet potato beverage on serum hepatic biomarker levels in healthy Caucasians with borderline hepatitis. *Eur. J. Clin. Nutr.*, *71(2)*, 290–292.

Mentha spicata

Falcone, P. H., Nieman, K. M., Tribby, A. C., *et al.* (2019) The attention-enhancing effects of spearmint extract supplementation in healthy men and women: A randomized, double-blind, placebo-controlled, parallel trial. *Nutr. Res.*, *64*, 24–38.

Kharchoufa, L., Bouhrim, M., Bencheikh, N., *et al.* (2021) Potential toxicity of medicinal plants inventoried in Northeastern Morocco: An ethnobotanical approach. *Plants*, *10(6)*, 1108.

Mahendran, G., Verma, S. K., Rahman, L. U. (2021) The traditional uses, phytochemistry and pharmacology of spearmint (*Mentha spicata* L.): A review. *J. Ethnopharmacol.*, *278*, 114266.

Ulbricht, C., Costa, D., M Grimes Serrano, J., *et al.* (2010) An evidence-based systematic review of spearmint by the natural standard research collaboration. *J. Diet. Suppl.*, *7(2)*, 179–215.

Zhang, L. L., Chen, Y., Li, Z. J., *et al.* (2022) Bioactive properties of the aromatic molecules of spearmint (*Mentha spicata* L.) essential oil: A review. *Food Funct.*, *13(6)*, 3110–3132.

Momordica charantia

Abdel-Rahman, R. F., Soliman, G. A., Saeedan, *et al.* (2019) Molecular and biochemical monitoring of the possible herb-drug interaction between *Momordica charantia* extract and glibenclamide in diabetic rats. *Saudi Pharm. J. 27(6)*, 803–816.

Cortez-Navarrete, M., Martínez-Abundis, E., Pérez-Rubio, K. G., *et al.* (2018) *Momordica charantia* administration improves insulin secretion in type 2 diabetes mellitus. *J. Med. Food*, *21(7)*, 672–677.

Lim, S. M., Sanip, Z., Ahmed Shokri, A., *et al.* (2018) The effects of *Momordica charantia* (bitter melon) supplementation in patients with primary knee osteoarthritis: A single-blinded, randomized controlled trial. *Complement. Ther. Clin. Pract.*, *32*, 181–186.

Zhang, F., Lin, L., Xie, J. (2016) A mini-review of chemical and biological properties of polysaccharides from *Momordica charantia*. *Int. J. Biol. Macromol.*, *92*, 246–253.

Zhao, Z. Z., Xiao, P. G. (2010) *Encyclopedia of Medicinal Plants, Volume 4*, World Publishing Corporation.

Morinda citrifolia

Leal-Silva, T., Souza, M. R., Cruz, L. L., *et al.* (2022) Toxicological effects of the *Morinda citrifolia* L. fruit extract on maternal reproduction and fetal development in rats. *Drug Chem. Toxicol.*, 1–7.

National Center for Complementary and Integrative Health (NCCIH) (n.d.) *Herbs at a glance: noni* [updated September 2016], NCCIH, Maryland. Available at https://nccih.nih.gov/health/noni (accessed 30 August 2018).

Torres, M. A. O., de Fátima Braga Magalhães, I., Mondêgo-Oliveira, R., *et al.* (2017) One plant, many uses: A review of the pharmacological applications of *Morinda citrifolia*. *Phytother. Res.*, *31(7)*, 971–979.

Wang, M. Y., Peng, L., Jensen, C. J., *et al.* (2013) Noni juice reduces lipid peroxidation-derived DNA adducts in heavy smokers. *Food Sci. Nutr.*, *1(2)*, 141–149.

West, B. J., Deng, S., Isami, F., *et al.* (2018) The potential health benefits of noni juice: A review of human intervention studies. *Foods*, *7(4)*, 58.

Moringa oleifera

Amaeze, O., Eng, H., Horlbogen, L., *et al.* (2021) Cytochrome P450 enzyme inhibition and herb-drug interaction potential of medicinal plant extracts used for management of diabetes in Nigeria. *Eur. J. Drug Metab. Pharmacokinet.*, *46(3)*, 437–450.

Attah, A. F., Akindele, O. O., Nnamani, P. O., *et al.* (2022) *Moringa oleifera* seed at the interface of food and medicine: Effect of extracts on some reproductive parameters, hepatic and renal histology. *Front. Pharmacol.*, *13*, 816498.

Chan Sun, M., Ruhomally, Z. B., Boojhawon, R., Neergheen-Bhujun, V. S. (2020) Consumption of *Moringa oleifera* Lam leaves lowers postprandial blood pressure. *J. Am. Coll. Nutr.*, *39(1)*, 54–62.

Ezzat, S. M., El Bishbishy, M. H., Aborehab, N. M., *et al.* (2020) Upregulation of MC4R and PPAR-α expression mediates the anti-obesity activity of *Moringa oleifera* Lam. in high-fat diet-induced obesity in rats. *J. Ethnopharmacol.*, *251*, 112541.

Zunica, E. R. M., Yang, S., Coulter, A., *et al.* (2021) *Moringa oleifera* seed extract concomitantly supplemented with chemotherapy worsens tumor progression in mice with triple negative breast cancer and obesity. *Nutrients*, *13(9)*, 2923.

Morus alba

Huh, H. W., Na, Y. G., Bang, K. H., *et al.* (2020) Extended intake of mulberry leaf extract delayed metformin elimination via inhibiting the organic cation transporter 2. *Pharmaceutics*, *12(1)*, 49.

Li, Y., Huang, L., Zeng, X., Zhong, G., *et al.* (2014) Down-regulation of P-gp expression and function after mulberroside A treatment: Potential role of protein kinase C and NF-kappa B. *Chem. Biol. Interact.*, *213*, 44–50.

Oliveira, A. M., Nascimento, M. F., Ferreira, M. R., *et al.* (2016) Evaluation of acute toxicity, genotoxicity and inhibitory effect on acute inflammation of an ethanol extract of *Morus alba* L. (Moraceae) in mice. *J. Ethnopharmacol.*, *194*, 162–168.

Thaipitakwong, T., Supasyndh, O., Rasmi, Y., Aramwit, P. (2020) A randomized controlled study of dose-finding, efficacy, and safety of mulberry leaves on glycemic profiles in obese persons with borderline diabetes. *Complement. Ther. Med.*, *49*, 102292.

Yuan, Q., Zhao, L. (2017) The mulberry (*Morus alba* L.) fruit a review of characteristic components and health benefits. *J. Agric. Food Chem.*, *65(48)*, 10383–10394.

Murraya koenigii

Abeysinghe, D. T., Alwis, D. D. D. H., Kumara, K. A. H., Chandrika, U. G. (2021) Nutritive importance and therapeutics uses of three different varieties (*Murraya koenigii*, *Micromelum minutum*, and *Clausena indica*) of curry leaves: An updated review. *Evid. Based Complementary Altern. Med.*, *2021*, 5523252.

Adebajo, A. C., Ayoola, O. F., Iwalewa, E. O., *et al.* (2006) Anti-trichomonal, biochemical and toxicological activities of methanolic extract and some carbazole alkaloids isolated from the leaves of *Murraya koenigii* growing in Nigeria. *Phytomedicine*, *13(4)*, 246–254.

Aniqa, A., Kaur, S., Sadwal, S. (2022) A review of the anti-cancer potential of *Murraya koenigii* (curry tree) and its active constituents. *Nutr. Cancer*, *74(1)*, 12–26.

Ocimum basilicum

Ahmadifard, M., Yarahmadi, S., Ardalan, A., *et al.* (2020) The efficacy of topical basil essential oil on relieving migraine headaches: A randomized triple-blind study. *Complement. Med. Res.*, *27(5)*, 310–318.

Alves Júnior, E. B., de Oliveira Formiga, R., de Lima Serafim, C. A., *et al.* (2020)

Estragole prevents gastric ulcers via cytoprotective, anti-oxidant and immunoregulatory mechanisms in animal models. *Biomed. Pharmacother. 130*, 110578.

Kalily, E., Hollander, A., Korin, B., *et al.* (2017) Adaptation of *Salmonella enterica* serovar Senftenberg to linalool and its association with antibiotic resistance and environmental persistence. *Appl. Environ. Microbiol., 83(10)*, e03398-16.

Nguyen, S., Huang, H., Foster, B. C., *et al.* (2014) Antimicrobial and P450 inhibitory properties of common functional foods. *World. J. Pharm. Pharm. Sci., 17(2)*, 254–265.

Sestili, P., Ismail, T., Calcabrini, C., *et al.* (2018) The potential effects of *Ocimum basilicum* on health: A review of pharmacological and toxicological studies. *Expert Opin. Drug Metab. Toxicol., 14(7)*, 679–692.

Ocimum tenuiflorum

Penmetsa, G. S., Pitta, S. R. (2019) Efficacy of *Ocimum sanctum, Aloe vera* and chlorhexidine mouthwash on gingivitis: A randomized controlled comparative clinical study. *Ayu, 40(1)*, 23–26.

Sarangi, S. C., Pattnaik, S. S., Joshi, D., *et al.* (2020) Adjuvant role of *Ocimum sanctum* hydroalcoholic extract with carbamazepine and phenytoin in experimental model of acute seizures. *Saudi Pharm. J., 28(11)*, 1440–1450.

Satapathy, S., Das, N., Bandyopadhyay, D., *et al.* (2017) Effect of tulsi (*Ocimum sanctum* Linn.) supplementation on metabolic parameters and liver enzymes in young overweight and obese subjects. *Indian J. Clin. Biochem., 32(3)*, 357–363.

Orthosiphon aristatus

Ashraf, K., Sultan, S., Adam, A. (2018) *Orthosiphon stamineus* Benth. is an outstanding food medicine: Review of phytochemical and pharmacological activities. *J. Pharm. Bioallied Sci., 10(3)*, 109–118.

Chung, Y. S., Choo, B. K. M., Ahmed, P. K., *et al.* (2020) A systematic review of the protective actions of cat's whiskers (misai kucing) on the central nervous system. *Front. Pharmacol., 11*, 692.

Pan, Y., Abd-Rashid, B. A., Ismail, Z., *et al.* (2011) *In vitro* effects of active constituents and extracts of *Orthosiphon stamineus* on the activities of three major human cDNA-expressed cytochrome P450 enzymes. *Chem. Biol. Interact., 190(1)*, 1–8.

Vogelgesang, B., Abdul-Malak, N., Reymermier, C., *et al.* (2011) On the effects of a plant extract of *Orthosiphon stamineus* on sebum-related skin imperfections. *Int. J. Cosmet. Sci., 33(1)*, 44–52.

Yehya, A. H. S., Subramaniam, A. V., Asif, M., *et al.* (2022) Anti-tumour activity and toxicological studies of combination treatment of *Orthosiphon stamineus* and gemcitabine on pancreatic xenograft model. *World J. Gastroenterol.*, *28(32)*, 4620–4634.

Pandanus amaryllifolius

Chiabchalard, A., Nooron, N. (2015) Antihyperglycemic effects of *Pandanus amaryllifolius* Roxb. leaf extract. *Pharmacogn. Mag.*, *11(41)*, 117–122.

Safriani, N., Rungkat, F. Z., Yuliana, N. D., Prangdimurti, E. (2021) Immuno-modulatory and anti-oxidant activities of select Indonesian vegetables, herbs, and spices on human lymphocytes. *Int. J. Food Sci.*, 6340476.

Tan, M. A., Ishikawa, H., An, S. S. A. (2022) *Pandanus amaryllifolius* exhibits *in vitro* anti-amyloidogenic activity and promotes neuroprotective effects in amyloid-β-induced SH-SY5Y cells. *Nutrients*, *14(19)*, 3962.

Pereskia bleo

Er, H. M., Cheng, E. H., Radhakrishnan, A. K. (2007) Anti-proliferative and mutagenic activities of aqueous and methanol extracts of leaves from Pereskia bleo (Kunth) DC (Cactaceae). *J. Ethnopharmacol.*, *113(3)*, 448–456.

Loo, S., Kam, A., Tam, J. P. (2022) Hyperstable EGF-like bleogen derived from cactus accelerates corneal healing in rats. *Front. Pharmacol.*, *13*, 942168.

Zareisedehizadeh, S., Tan, C. H., Koh, H. L. (2014) A review of botanical characteristics, traditional usage, chemical components, pharmacological

activities, and safety of *Pereskia bleo* (Kunth) DC. *Evid. Based Complementary Altern. Med.*, *2014*, 326107.

Plectranthus amboinicus

Leu, W. J., Chen, J. C., Guh, J. H. (2019) Extract from *Plectranthus amboinicus* inhibit maturation and release of interleukin 1β through inhibition of NF-κB nuclear translocation and NLRP3 inflammasome activation. *Front. Pharmacol.*, *10*, 573.

Lukhoba, C. W., Simmonds, M. S., Paton, A. J. (2006) *Plectranthus*: A review of ethnobotanical uses. *J. Ethnopharmacol.*, *103(1)*, 1–24.

Ramli, N., Ahamed, P. O., Elhady, H. M., Taher, M. (2014) Antimalarial activity of Malaysian *Plectranthus amboinicus* against *Plasmodium berghei*. *Pharmacognosy Res*, *6(4)*, 280–284.

Punica granatum

Ali, N., Jamil, A., Shah, S. W., *et al.* (2017) Spasmogenic and spasmolytic activity of rind of *Punica granatum* Linn. *BMC Complement. Altern. Med.*, *17(1)*, 97.

Awad, R., Mallah, E., Khawaja, B. A., *et al.* (2016) Pomegranate and licorice juices modulate metformin pharmacokinetics in rats. *Neuro Endocrinol. Lett.*, *37(3)*, 202–206.

Grabež, M., Škrbić, R., Stojiljković, M. P., *et al.* (2022) A prospective, randomized, double-blind, placebo-controlled trial of polyphenols on the outcomes of inflammatory factors and oxidative stress in patients with type 2 diabetes mellitus. *Rev. Cardiovasc. Med.*, *23(2)*, 57.

Kojadinovic, M., Glibetic, M., Vucic, V., *et al.* (2021) Short-term consumption of pomegranate juice alleviates some metabolic disturbances in overweight patients with dyslipidemia. *J. Med. Food*, *24(9)*, 925–933.

Rosmarinus officinalis

Fernández, L. F., Palomino, O. M., Frutos, G. (2014) Effectiveness of *Rosmarinus officinalis* essential oil as anti-hypotensive agent in primary hypotensive patients and its influence on health-related quality of life. *J. Ethnopharmacol.*, *151(1)*, 509–516.

Lin, K. I., Lin, C. C., Kuo, S. M., *et al.* (2018) Carnosic acid impedes cell growth and enhances anticancer effects of carmustine and lomustine in melanoma. *Biosci. Rep.*, *38(4)*, BSR20180005.

Nasiri, A., Boroomand, M. M. (2021) The effect of rosemary essential oil inhalation on sleepiness and alertness of shift-working nurses: A randomized, controlled field trial. *Complement. Ther. Clin. Pract.*, *43*, 101326.

Nematolahi, P., Mehrabani, M., Karami-Mohajeri, S., Dabaghzadeh, F. (2018) Effects of *Rosmarinus officinalis* L. on memory performance, anxiety, depression, and sleep quality in university students: A randomized clinical trial. *Complement. Ther. Clin. Pract.*, *30*, 24–28.

Phipps, K. R., Lozon, D., Baldwin, N. (2021) Genotoxicity and subchronic toxicity studies of supercritical carbon dioxide and acetone extracts of rosemary. *Regul. Toxicol. Pharmacol.*, *119*, 104826.

Strobilanthes crispa

Lim, K. T., Lim, V., Chin, J. H. (2012) Subacute oral toxicity study of ethanolic leaves extracts of *Strobilanthes crispus* in rats. *Asian Pac. J. Trop. Biomed.*, *2(12)*, 948–952.

Ng, M. G., Ng, C. H., Ng, K. Y., *et al.* (2021) Anticancer properties of *Strobilanthes crispus*: A review. *Processes*, *9(8)*, 1370.

Yong, Y. F., Liew, M. W. O., Yaacob, N. S. (2022) Inhibition of cytochrome P450s by *Strobilanthes crispus* subfraction (F3): Implication for herb-drug interaction. *Eur. J. Drug Metab. Pharmacokinet.*, *47(3)*, 431–440.

Vernonia amygdalina

Amaeze, O., Eng, H., Horlbogen, L., *et al.* (2021) Cytochrome P450 enzyme inhibition and herb-drug interaction potential of medicinal plant extracts used for management of diabetes in Nigeria. *Eur. J. Drug Metab. Pharmacokinet.*, *46(3)*, 437–450.

Challand, S., Willcox, M. (2009) A clinical trial of the traditional medicine *Vernonia amygdalina* in the treatment of uncomplicated malaria. *J. Tradit. Complement. Med.*, *15(11)*, 1231–1237.

Momoh, M. A., Muhamed, U., Agboke, A. A., *et al.* (2012) Immunological effect of aqueous extract of *Vernonia amygdalina* and a known immune booster called Immunace® and their admixtures on HIV/AIDS clients: A comparative study. *Asian Pac. J. Trop. Biomed.*, *2(3)*, 181–184.

Njan, A. A., Adzu, B., Agaba, A. G., *et al.* (2008) The analgesic and antiplasmodial activities and toxicology of *Vernonia amygdalina*. *J. Med. Food*, *11(3)*, 574–581.

Vitex trifolia

Hung, N. H., Quan, P. M., Satyal, P., *et al.* (2022) Acetylcholinesterase inhibitory activities of essential oils from Vietnamese traditional medicinal plants. *Molecules*, *27(20)*, 7092.

Siew, Y. Y., Yew, H. C., Neo, S. Y., *et al.* (2019) Evaluation of anti-proliferative activity of medicinal plants used in Asian Traditional Medicine to treat cancer. *J. Ethnopharmacol.*, *235*, 75–87.

Wee, H. N., Neo, S. Y., Singh, D., *et al.* (2020) Effects of *Vitex trifolia* L. leaf extracts and phytoconstituents on cytokine production in human U937 macrophages. *BMC Complement. Altern. Med.*, *20(1)*, 91.

Yan, C. X., Wei, Y. W., Li, H., *et al.* (2023) *Vitex rotundifolia* L. f. and *Vitex trifolia* L.: A review on their traditional medicine, phytochemistry, pharmacology. *J. Ethnopharmacol.*, *308*, 116273.